Marcello Spinella, PhD

Concise Handbook of Psychoactive Herbs

Medicinal Herbs for Treating Psychological and Neurological Problems

Pre-publication REVIEWS, COMMENTARIES, EVALUATIONS . . .

"This book is a much-needed text that provides the reader with the essential background about psychoactive herbs. It is an outstanding summary that covers commonly used herbal supplements such as ginseng as well as the more exotic plant-based options such as hallucinogenic mushrooms. The book will be a valuable reference text for herbalists, physicians, researchers, policymakers, and all other professionals with an interest in this area."

Edzard Ernst, MD, PhD, FRCP
Professor of Complementary Medicine, Peninsula Medical School, Exeter, Devon, United Kingdom

"This well-written book provides an informative overview of the topic. It offers an unbiased view of the current state of knowledge of various herbs and their potential psychoactive properties. What makes for an enjoyable read is the integration of the historical perspective, cultural perspective, and review of current clinical perspectives. Dr. Spinella clearly differentiates folklore from proven clinical fact, highlighting the importance of well-designed clinical trials to determine effectiveness of any compound thought to affect mood and behavior. This book is a scientifically sound introduction to the role of herbs in mental health."

David A. Baron, DO, MSEd
Professor and Chair, Department of Psychiatry, Temple University School of Medicine

More pre-publication
REVIEWS, COMMENTARIES, EVALUATIONS . . .

"This book will be helpful to health care practitioners and patients who are not well versed in the therapeutic utility and potential risks associated with herbal preparations. Until now, the average person interested in the psychopharmacology of herbals did not have access to an easily understandable and reputable source of information. Health care practitioners will be better prepared to provide intelligent advice to patients who take over-the-counter herbal medications."

Juan Sanchez-Ramos, MD, PhD
Professor of Neurology,
Psychiatry, and Pharmacology,
University of South Florida,
Tampa

The Haworth Herbal Press®
The Haworth Reference Press™
Imprints of The Haworth Press, Inc.
New York • London • Oxford

NOTES FOR PROFESSIONAL LIBRARIANS AND LIBRARY USERS

This is an original book title published by The Haworth Herbal Press© and The Haworth Reference Press™, imprints of The Haworth Press, Inc. Unless otherwise noted in specific chapters with attribution, materials in this book have not been previously published elsewhere in any format or language.

CONSERVATION AND PRESERVATION NOTES

All books published by The Haworth Press, Inc. and its imprints are printed on certified pH neutral, acid-free book grade paper. This paper meets the minimum requirements of American National Standard for Information Sciences-Permanence of Paper for Printed Material, ANSI Z39.48-1984.

Concise Handbook of Psychoactive Herbs

Medicinal Herbs for Treating Psychological and Neurological Problems

THE HAWORTH PRESS
Titles of Related Interest

The ABC Guide to Herbs by Mark Blumenthal, Josef Brinckmann, Kara Dinda, Alicia Goldberg, and Bernd Wollschlaeger

Botanical Medicines: The Desk Reference for Major Herbal Supplements, Second Edition by Dennis J. McKenna, Kenneth Jones, and Kerry Hughes

The Healing Power of Chinese Herbs and Medicinal Recipes by Joseph P. Hou and Youyu Jin

Medicinal Herbs: A Compendium by Beatrice Gehrmann, Wolf-Gerald Koch, Claus O. Tschirch, and Helmut Brinkmann

Non-Timber Forest Products: Medicinal Herbs, Fungi, Edible Fruits and Nuts, and Other Natural Products from the Forest edited by Marla R. Emery and Rebecca J. McLain

Tyler's Herbs of Choice: The Therapeutic Use of Phytomedicinals by James E. Robbers and Varro E. Tyler

Understanding Medicinal Plants: Their Chemistry and Therapeutic Action by Bryan A. Hanson

Concise Handbook of Psychoactive Herbs

Medicinal Herbs for Treating Psychological and Neurological Problems

Marcello Spinella, PhD

The Haworth Herbal Press®
The Haworth Reference Press™
Imprints of The Haworth Press, Inc.
New York • London • Oxford

The Haworth Press, Inc., 10 Alice Street, Binghamton, NY 13904-1580.

PUBLISHER'S NOTE
This book has been published solely for educational purposes and is not intended to substitute for the medical advice of a treating physician. Medicine is an ever-changing science. As new research and clinical experience broaden our knowledge, changes in treatment may be required. While many potential treatment options are made herein, some or all of the options may not be applicable to a particular individual. Therefore, the author, editor, and publisher do not accept responsibility in the event of negative consequences incurred as a result of the information presented in this book. We do not claim that this information is necessarily accurate by the rigid scientific and regulatory standards applied for medical treatment. **No warranty, expressed or implied, is furnished with respect to the material contained in this book. The reader is urged to consult with his/her personal physician with respect to the treatment of any medical condition.**

Cover design by Kerry E. Mack.

Library of Congress Cataloging-in-Publication Data

Spinella, Marcello.
Concise handbook of psychoactive herbs : medicinal herbs for treating psychological and neurological problems / Marcello Spinella.
p. cm.
Includes bibliographical references.
ISBN 0-7890-1857-8 (hard : alk. paper)—ISBN 0-7890-1858-6 (soft : alk. paper)
1. Psychotropic plants—Handbooks, manuals, etc. 2. Psychopharmacology—Handbooks, manuals, etc. 3. Herbs—Therapeutic use—Handbooks, manuals, etc.
[DNLM: 1. Psychotropic Drugs—pharmacology—Handbooks. 2. Phytotherapy—Handbooks. 3. Plants, Medicinal—Handbooks. 4. Psychotropic Drugs—therapeutic use—Handbooks. 5. Substance-Related Disorders—drug therapy—Handbooks. QV 39 S757c 2004] I.Title.
RM315.S655 2004
615'.78—dc22

2004006233

CONTENTS

ABOUT THE AUTHOR

Marcello Spinella, PhD, is Associate Professor of Psychology at the Richard Stockton College of New Jersey in Pomona. Dr. Spinella received his PhD in neuropsychology from the City University of New York. His psychology internship was at the Kessler Institute for Rehabilitation and his postdoctoral fellowship was at the Center for Head Injuries and the New Jersey Neuroscience Institute at JFK Medical Center. His interests include physiological psychology, clinical neuropsychology, psychometrics, psychopharmacology, herbal medicines, neural basis of cognition, and aging and mental health.

Dr. Spinella is the author of *The Psychopharmacology of Herbal Medicines: Plant Drugs That Alter Mind, Brain, and Behavior.* He has written articles for such journals as *Pharmacology, Biochemistry and Behavior,* the *Journal of Neurology, Neurosurgery, and Psychology, Physiology and Behavior, Analgesia,* and *Pain*. He is a member of the New Jersey Neuropsychological Society, the New York Neuropsychology Group, and the Society for Neuroscience.

Chapter 1

Introduction

The use of plants for their psychological effects is universal among humans. People of almost all cultures and throughout human history have used them routinely. The effects of such plants have captured our interest, and knowledge about psychoactive plants acquired by our ancestors has been handed down across generations. This traditional knowledge is now encountering modern science, allowing us to examine plants in a way that was not previously possible. What chemicals the plants are composed of, how they access the brain, and how they alter the mind are questions we are now actively answering.

This merger of the ancient and the modern is an ongoing process, and connections between the two are made painstakingly, one step at a time. An enormous amount of knowledge about plants is available, and the scientific aspects can be very technical. However, despite this seemingly intimidating exterior, the whole subject is surprisingly comprehensible. When expressed in lay terms, it suddenly becomes familiar. Most of us use **psychoactive** drugs for one purpose or another. If you don't, then you certainly know people who do. When the technical veneer is lifted, much of their use makes intuitive sense.

PLANTS AS MEDICINES

Plants are the oldest source of medicines, and probably still the main source. Before the modern pharmaceutical industry existed, people relied on folk knowledge and apothecaries. Only very recently have we become able to circumvent plants and synthesize **drugs** without them. Even though we are no longer completely dependent on plants to make drugs for us, we still resort to them much of the time. The painkillers codeine and morphine are still extracted from the **opium poppy** because to chemically synthesize them is difficult and expensive.

If we consider the nonmedicinal use of psychoactive medicines, it becomes obvious that we still get most of our drugs from plants. Consider all the caffeine that is consumed on a daily basis in the form of coffee and tea—it's the most common stimulant in the world. Add to that the tobacco that is smoked or chewed and the cocaine that is snorted or smoked. From these examples, we see how long the list will become.

Is It a Drug or a Nutrient?

It's important to make a distinction between drugs and nutrients. Both can come from plants, and both alter our bodily functions. The distinguishing factor is that a **nutrient** is essential for life, while a drug is not. If a person is deprived of vitamin C, illness and death will follow. On the other hand, no one ever died from a lack of aspirin. People have lived entire lifetimes without ever taking Valium (although they may have been a little more nervous). A poison, in contrast, is a chemical that alters body functions for the worse, causing illness or death. In reality, any drug can be toxic if used in the wrong way. A physician in the 1500s named Paracelsus astutely noted that the difference between a remedy and a poison is a matter of dosage.

As a word of caution, the way we prepare vitamin supplements blurs the drug-nutrient distinction. For most of human history, nutrients were derived exclusively from food. Whatever vitamins, minerals, amino acids, and so on that we needed were found in what we ate. Now we are able to refine individual nutrients, which allows us to take them alone and in large amounts, more than would be physically possible from eating food. In this sense, nutrients are more like drugs, which has prompted the creation of the term **nutraceutical.**

Herbal versus Synthetic Drugs

Herbal medicines are drugs in every sense of the word. As such, they can help or harm, depending on how they are used. However, there are a few general differences between herbal and **pharmaceutical** drugs. Herbal drugs tend to be more dilute than pharmaceutical drugs (see Table 1.1). The amount of caffeine in a 200 mg over-the-counter tablet requires about 20 g of coffee beans. Although aspirin

TABLE 1.1. Differences between herbal and pharmaceutical drugs.

Factor	Herbal	Pharmaceutical
Concentration	More dilute	More concentrated
Administration instructions	Variable	Prescribed
Number of drugs contained	Usually multiple	Usually a single drug

was originally derived from willow bark, the amount of willow bark tea one would need to drink to get the equivalent of a few aspirins would make one ill. That doesn't mean herbal medicines are weaker, just less concentrated.

Another major difference is the way drugs are taken, which makes an enormous difference in the effects produced. Pharmaceutical drugs are designed to be taken in a certain way, which makes the outcome more predictable when the instructions are followed. For example, Advil tablets are designed to be swallowed, and the nicotine patch is designed to be attached to the skin. Herbal medicines, you may have noticed, don't necessarily have a standard way to be administered. The same herb may be found in many different forms (Figure 1.1).

The simplest way to consume an herbal medicine is to eat it. However, in many cases the plant tastes bad or is too bulky, making this impractical. The next step would be to dry and grind the herb, making it easier to swallow in capsules. Both of these methods use the entire plant. Alternately, a tea (or infusion) could be made, which is easy to drink, but it includes only the chemicals that seep into the water. Some important chemicals dissolve well in fat but not in water, and these would be left out of a tea.

A further step in preparing an herbal medicine is making an extract. This is done by chopping up a plant and soaking it in a solvent (such as alcohol or acetone). The active chemicals seep into the extract, and the plant matter is removed. The solvent is then allowed to evaporate, leaving a powder of concentrated chemicals from the plant. The advantage of an extract is that it reduces the amount that needs to be consumed. One tablet of herbal extract can be equivalent to a bale of dried weeds. Another advantage is that extracts can be standardized, which means the amounts of active chemicals can be ensured. This allows control of the dosage of the herbal drug. Otherwise, it would be like taking a Tylenol tablet from a bottle labeled,

No processing	Raw plant
↑	Dried and ground, encapsulated
	Infusion (tea)
	Extract
↓	Single drug refined out
Maximum processing	Modification of the refined drug

FIGURE 1.1. Ways herbs are prepared and consumed.

"Each tablet contains somewhere between 10 and 1,000 mg. We're not really sure." Nobody would buy a medicine labeled that way, but that is literally what happens when we buy nonstandardized herbal medicines. To be fair, an exact dosage is not always absolutely necessary: a cup of chamomile tea to soothe the stomach doesn't require precision dosing. But in many other cases, a standardized herb is useful or even necessary.

Another procedure in processing an herbal medication is to refine out a single drug from the extract. Opium contains more than thirty drugs, but we selectively refine out morphine and codeine. Sometimes only a certain drug is needed, rather than all of them. Morphine alone is an effective painkiller, but the other drugs in opium would cause unwanted medical complications if doctors prescribed all of them together. A further step would be to chemically modify the drug that was refined. For example, morphine refined from opium is modified to make heroin.

A third major difference between herbal and pharmaceutical drugs is that a pharmaceutical drug is usually a single drug. Herbal medications, on the other hand, usually contain multiple active drugs. As mentioned previously, opium is a virtual medicine cabinet, containing more than thirty drugs, each with different effects. Taking an herbal medicine encompasses the combined effects of all of the drugs in it. It is difficult to know all of the effects of a single drug; to know the effects of different drugs interacting with one another is much more complex. Sometimes it's more advantageous to prescribe a single refined drug from the plant, while in other cases the combination is better.

THE USE OF HERBAL MEDICINES

The use of complementary and alternative medicine (CAM) is very common. This blanket term includes many different therapies including herbal medicines, acupuncture, massage therapy, and so on. Surveys indicate that approximately 40 percent of Americans use one or more forms of CAM and that this percentage has been increasing over the past few years. People use CAM especially to treat chronic conditions, such as back problems, anxiety, depression, and headaches. This is not surprising, since these conditions often slip through the cracks of orthodox medicine. The estimated cost of all these therapies is more than $20 billion. Whether or not one agrees with the use of CAM, it's a significant force.

The Appeal of Herbal Medicines

People use CAM for many reasons. Many people have the perception that CAM therapies, including herbal medicines, are more "natural" and have greater aesthetic appeal than synthetic drugs. Using a plant to heal an illness is a much more appealing idea than taking an expensive, exotically named chemical that presumably came from a factory. However, a drug has no intrinsic benefits simply because it comes from a plant. Although it may be better to use an herbal remedy in some instances, it is not necessarily the case. Herbal medicines have great potential for harm when improperly used, and they have the added disadvantage that research to guide proper usage is lacking for many of them. We're still learning about them. Thus, all of the potential risks must be weighed against the collective benefits.

Another reason for the appeal of herbal medicines and CAM is that they provide more options and autonomy. When someone is confronted with an illness, and the options provided by modern medicine are unappealing (e.g., painful and/or expensive treatments, limited chances for success), CAM provides other options. Whether they actually work is a separate matter, but having more options is psychologically appealing.

CAM also allows people more autonomy. Modern medicine has advanced with science, meaning that understanding and treating an illness requires a great deal of training in the form of medical school. This often creates a greater gap between the doctor and the patient, who must passively follow the physician's instructions for lack of technical knowledge. However, seeking other forms of treatment, whether accepted or not, puts the patient in a proactive role. Psychological research consistently shows that helplessness leads to distress, and that being autonomous and proactive protects against it.

This does not address whether CAM therapies work. Some may work very well and some may not work at all. Nonetheless, these factors serve as incentives to use CAM.

Herbal Medicine in the Medical Field

Some estimates suggest that only 40 percent of people tell their physicians that they are using CAM therapies. Millions also take prescription medications along with herbal medications and/or high doses of vitamins. This creates a great potential for dangerous interactions. Many cases have been documented of harmful interactions between herbal medicines and prescribed or over-the-counter pharmaceutical drugs. For everyone's benefit, physicians need to make patients more comfortable about disclosing their use of CAM, and patients need to be more forthcoming with their physicians. Many drug-herb interactions are not yet known, but at least the known interactions can be avoided.

Physicians prescribing herbal medicines is more common in Europe and Asia. The American medical system is showing some signs of change, however, and many medical schools are adding courses on CAM. Medical journals are now routinely publishing articles about the proper and improper uses of herbal medicines.

However, the traditional reluctance of many physicians to prescribe herbal medications is understandable. Physicians are obligated to treat patients according to a standard of care, i.e., the most effective and reliable treatments at their disposal should be used to diagnose and treat illnesses. Since limited research has been done on the safety and effectiveness of many herbal medicines, much of the hesitancy of the medical community is well intentioned. What is needed at this point is increased research and education on such therapies, to guide both physicians and the public.

Government Regulation of Herbal Medicines

Several laws have been passed to control the way psychoactive plant medications are sold and used. Before the twentieth century, few if any laws regulated psychoactive drugs. At the beginning of the twentieth century, cocaine and heroin could be purchased at a pharmacy, and patent medicine salesmen could make whatever claims they wished for their products. Clearly the situation had to change. Since then, several laws have been enacted to address the safety, effectiveness, and abuse potential of legally marketed drugs.

Three main organizations in the United States regulate the use of psychoactive plant drugs: the Food and Drug Administration (FDA), the Drug Enforcement Agency (DEA), and the Bureau of Alcohol, Tobacco, and Firearms (ATF). The FDA in general regulates legal drugs, such as prescriptions, over-the-counter medications, and dietary supplements. According to the Dietary Supplement Health and Education Act, which was passed by the U.S. Congress in 1994, herbal medicines are regulated as dietary supplements and not as drugs, providing they don't make any specific claims to treat a disease. So, for example, one cannot sell St. John's wort claiming it "cures depression." Instead, only a suggestive statement such as "improves mood" can be used. In order to sell a drug making specific claims about treating an illness, the drug must go through the FDA's drug approval process (see Box 1.1). This is an extensive and costly process, taking twelve or more years and costing $350 to $500 mil-

BOX 1.1. The Drug Approval Process

1. Preclinical trials
   - Last about five years
   - Cell culture and animal research to determine the effects of a new drug
2. Investigational new drug (IND) application is filed with the FDA
3. Clinical trials
   - Last about six years (total)
   - Phase I—Preliminary research in healthy volunteers (20-80)
   - Phase II—Testing in a limited number of patients (100-300)
   - Phase III—Large-scale testing in patients (1,000 or more)
4. New drug application (NDA) is filed with the FDA
   - Takes about 2.5 years
5. Phase IV—Postmarketing surveillance
   - The approved drug is marketed
   - Continued research and surveillance are done to monitor drug safety
   - About 20 percent of all new drugs are eventually found to have serious adverse effects

lion. Since plants cannot be patented, a private company has no incentive to expend that much time and resources to approve a drug that anyone will be able to sell. Consequently, little financial incentive exists to research herbal medicines.

The DEA is a law enforcement organization that regulates illegal drugs and prescription drugs with abuse potential. The Controlled Substances Act, which was passed in 1970, classifies drugs according to their currently accepted medical uses and potential for abuse or addiction (see Box 1.2). Some of the designations of these drugs are debated. For example, cannabis (marijuana) is listed under Schedule I, meaning that it has no accepted medical value. This is a highly debated matter, and research suggests that it has legitimate medical uses.

The ATF is a branch of the U.S. Treasury Department that regulates the sale of alcohol and tobacco products. It is also a law enforce-

BOX 1.2. The Controlled Substances Act

Schedule I	The drug has no currently accepted medical uses. The drug is believed to have high potential for abuse. Examples: heroin, LSD, and marijuana
Schedule II	Drug has accepted medical uses, with severe restrictions. Abuse potential is high. Examples: morphine, PCP, cocaine, methadone, and methamphetamine
Schedule III	Drug has accepted medical uses. Abuse potential is lower than for Schedule I and II drugs. Examples: Anabolic steroids, codeine, some barbiturates
Schedule IV	Drug has accepted medical uses. Abuse potential is lower than for Schedule III drugs. Examples: Darvon, Talwin, Equanil, Valium, and Xanax
Schedule V	Drug has accepted medical uses. Abuse potential is lower than for Schedule IV drugs. Examples: Buprenorphine, pyrovalerone, low doses of codeine

Source: The Controlled Substances Act [CSA], Title II of the Comprehensive Drug Abuse Prevention and Control Act of 1970.

ment agency that polices the legal production and sale of alcohol and tobacco products.

WHY DO PLANTS MAKE DRUGS?

On the surface, it seems unusually convenient that plants are able to make medicines for humans. To understand this requires a slight shift in perspective. Charles Darwin eloquently established that evolution occurs by natural selection. Organisms possessing the most adaptive traits, allowing them to survive and reproduce, will likely pass those traits to their offspring. Green, multicellular plants evolved at approximately the same time as insects, who found an abundant source of food in the plants. Although this was good news for the insects, it was bad news for the plants. In order to thrive, then, plants developed defenses throughout generations. This process is blind, working mostly through serendipity and fortunate accidents. Many plants developed spines or cuticles, making it difficult to manipulate or eat them. Some developed chemical defenses, their own insecticides. Of course, the plant predators counteradapted over generations, and those resistant to the toxic plant chemicals again had an abundant source of food.

It is from this ongoing chemical warfare that our drugs derive. Nicotine, for example, is highly toxic to most insects and ensured the survival of the tobacco plant. Caffeine is also toxic, protecting the coffee plant. Ironically, caffeine is so toxic that the plants end up poisoning themselves after many years, but only after they have produced many offspring.

HOW HUMANS CAME TO USE HERBAL MEDICINES

At some time in human history, we acquired the use of herbs for medicinal purposes. Most likely, these uses were serendipitous discoveries in our searches for edible plants. In the course of trying different plants, people might accidentally consume a plant that alleviated a particular ill. No forethought was necessary, only trial-and-error learning. One of the basic laws of psychology is behavioral reinforcement, as described by Edwin Thorndike around the turn of the twentieth century: any action followed by a pleasurable consequence is likely to be repeated, and any action followed by aversive consequences is less likely to be repeated. So, for example, a person eating ginger may have noticed that his or her nausea subsided. Thus, in the future, that individual would be more likely to seek it out when feeling queasy.

As humans possess language, we are able to give names to the plants we discover. This allows knowledge about their uses to accumulate and be passed down through generations. Each culture seems to have amassed numerous plants that are taken as folk remedies for specific illnesses. Each culture may have invented systems for explaining how these plants heal. In Asian systems, herbal medicine is based on the theory of balancing yin and yang energy. In Europe, a belief called the **doctrine of signatures** was created, which held that the shape or color of a plant indicates its therapeutic uses. (For example, if a plant was red, it would be used to treat blood-related disorders.) Whether these explanations are accurate is not highly important; if people discover the right use for the right plant, the practice is bound to continue.

To further emphasize this point, a group of chimpanzees in Tanzania has been found to be using a plant called *Aspilia* in a medicinal fashion. The chimps were observed swallowing the leaves whole without chewing them, so apparently they are not eating them for the taste. They derive no apparent nutrition from the plant, so the question is, why are they eating them? The answer lies in a chemical contained in the plant called thiarubine-A, which is effective against parasites and microorganisms of the digestive tract. Chimps who eat the plant seem to be free of these kinds of problems. Presumably, the chimps have not done any biochemical research or any FDA-approved clinical trials on the *Aspilia* plant. Nonetheless, eating the plant apparently makes them feel noticeably better. If the benefit occurs reliably, and close enough in time to the behavior (i.e., eating the plant), reinforcement occurs and the practice continues. It's the same for animals as for humans.

THE SCIENCE OF HERBAL MEDICINES

Clinical Drug Trials of Herbal Medicines

We now seek proof that herbal medicines work. Whether the uses are handed down from tradition or are new uses conceived recently, it's useful to know if an herb is effective, otherwise, it may not be worth the time and effort required to obtain and use it. After all, growing and preparing herbs costs money.

A clinical trial for a drug is essentially an experiment and must have certain elements to be worthwhile (see Box 1.3). All drug trials, especially those involving psychoactive drugs, must compare the real drug to a **placebo,** which is an inactive treatment. Placebo effects, also called expectancy effects, result from a person's expectations rather than from the treatment itself. For example, many people given an antidepressant drug start to feel a lift in their mood due to the expectation that the drug will help, even though the drug takes several weeks to actually work. So any drug study must have at least two groups of subjects: one getting the real drug, and another getting an indistinguishable but inactive placebo. Another phenomenon, **nocebo,** also can occur due to one's expectations. Whereas placebo effects are imagined benefits from a drug, nocebo effects are imagined side effects. If a plain glass of water is given to 100 people, a number of them will experience symptoms (e.g., headache, upset stomach, dizziness) and attribute it to the water. To rule this out, drugs are compared to the placebo (or nocebo) to see if the drug really produces any additional side effects.

Second, subjects must be assigned to the drug or placebo groups randomly. This removes the possibility that the researcher could influence the outcome, whether consciously or unconsciously. Third, there must be a double blind. This means that neither the researcher nor the subjects know who is getting the placebo or the real drug until the experiment is over. So, a third party is recruited to set up the drugs and label them, keeping their real identity secret until after the study is completed.

Only when these guidelines are closely followed can a study yield firm conclusions about how well a drug works. Although not neces-

BOX 1.3. Elements of a Clinical Drug Trial

Necessary Elements	Other Useful Elements
Compare the drug to a placebo	Dose-response relationship
Randomly assign subjects the drug or placebo	Comparison to an existing, established treatment
Double blind	

sary, it is often useful to show a dose-response relationship. By using different dosages of the drug, one can show that the intensity of effect varies with how much of the drug is given. Again, this strengthens the case that the effects are due to the drug and not imagined, since the subjects do not know what dose they are getting or whether they are even getting the real drug.

In some cases, it's unethical to give a placebo. This is especially true when the drug is being tested to treat a serious or life-threatening illness. Giving a placebo could potentially kill the person, which is not an acceptable risk. In these cases, a new drug could be compared to an existing, established drug. Rather than showing that the new drug is better than placebo, one can look at whether it is as good or maybe even better than the drug already being used.

Scientific Disciplines in Herbal Drugs

Several branches of science help us understand herbal medicines. Naturally, **botany** is an important starting point since this branch of science deals with plant life. Next, **biochemistry** tells us what the plant is composed of and what its active chemicals may be. **Pharmacology** is the study of drugs, including how they affect the body and how the body affects them. **Psychopharmacology** is the branch of pharmacology that specifically studies how drugs affect the mind, brain, and behavior. Other related fields are **ethnobotany** and **ethnopharmacology,** which study the use of plants and drugs in the context of cultures. Culture has a powerful influence on how plants and drugs are used, and what we expect to happen when they are used. For example, consider how much education you've had regarding alcohol. Most people haven't formally studied it, yet the average person knows a lot about it. We have informally gained knowledge about the effects of alcohol, how much a person must consume to get those effects, and when it's considered acceptable or unacceptable to use alcohol (e.g., drinking moderately on a weekend in the company of friends is okay, while getting drunk on a weekday morning is generally frowned upon).

Since psychoactive plants produce their effects by altering the nervous system, the **neurosciences** are important to understand how they work. Knowing the structure, function, and chemical composition of the brain lets us understand how drugs can affect them. Finally, the clinical disciplines of **psychiatry, psychology, neurology,** and **neuropsychology** tell us about illnesses of the mind and brain. Many psychoactive herbal medicines have been used in an attempt to treat these illnesses, in some cases with a fair degree of success.

SUMMARY

Plant medicines are clearly a part of human life, and probably have been for as long as we have existed. Further, humans take an active interest in psychoactive plants for a variety of reasons, including therapeutic, recreational, and religious purposes.

Some drug treatments have been thoroughly researched and are now part of mainstream medical treatment. Others have not undergone enough formal research and are still considered "alternative medicine." The goal of this book is to summarize and explain the use of plants for mind-altering purposes. Knowing how they work will make a difference in how we use them. Knowing whether or not they really do work makes a difference in our decision whether to use them.

Chapter 2

How the Brain Works

A basic understanding of the nervous system helps us understand the effects of psychoactive drugs. It gives a context for understanding their effects, since, by definition, psychoactive drugs work their magic by altering the function of the brain. A new drug's effects can be fairly accurately predicted based on what it does in the brain.

MIND, BRAIN, AND BEHAVIOR

Before delving into the nervous system, it's important to gain a perspective on the mind, brain, and behavior. All three of these are essentially different aspects of the same thing. The brain (or entire nervous system) is the physical organ in this triad: the cells, chemicals, and electrical impulses that underlie the whole thing. The mind is the personal, subjective aspect. It is the internal and private experience that each person has. This includes our thoughts, emotions, sensory perceptions, memories, and motivations. One person's mind cannot be directly known by another person. Instead we must infer what's in another person's mind by his or her behavior: actions, facial expressions, and speech. These things are outwardly observable. Our ability to infer another's state of mind and communicate it is so well developed that it's easy to forget that we are even making an inference. Of course, it's not perfect and we are sometimes wrong. Everyone knows the experience of thinking that another person is sad, angry, and so on, and later finding out that this is untrue.

Cells of the Brain

The basic unit of the nervous system is the **neuron.** It's the type of cell that processes information in the brain, whereas the other cells serve more of a support function (see Figure 2.1). Despite their amaz-

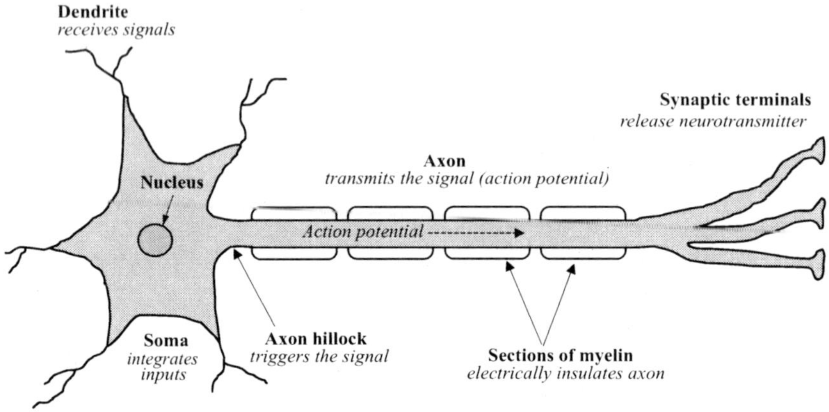

FIGURE 2.1. The neuron.

ing function, they are very much like other cells in the body. Their main difference is that they pass electrical-chemical messages among themselves, and they are connected in large, complex networks. Linked to our senses, they register what is happening in the world around us, and linked to our muscles they allow us to move and act. The adult human brain is estimated to have approximately 100 billion neurons. Keep in mind that this is an estimate, since it's understandably hard to count them.

A neuron has certain core components. It has a protein skeleton that holds its shape. In some diseases, such as **Alzheimer's disease,** these proteins become tangled and the neurons can't function properly. The outer "skin" of a neuron is called the cell membrane. It is made of fluid, fatty material. Some proteins are stuck into this membrane, which means they have access to the outside of the neuron. Since the membrane is fatty, chemicals that dissolve well in fat can pass right through it (such as the hormones testosterone and estrogen). Neurons also have many organelles, analogous to miniature internal organs. The **nucleus** is an organelle that contains all of our DNA, or a complete genetic blueprint. From this blueprint, the neuron can make anything that it needs, i.e., all of the proteins and enzymes. **Mitochondria** are other organelles that are needed to make energy. They take oxygen and the sugar glucose (delivered by the blood supply to the brain) and convert them into ATP (adenosine triphosphate). ATP is the universal form of energy used in cells.

Parts of the Neuron

Dendrites

Dendrites (from the Greek word *dendros,* meaning branches of a tree) extend from the body of a neuron. These branches are typically the receiving end, taking in messages from other neurons. Dendrites even have branches on their branches, which allows more space for connections. In fact, whenever the brain learns new information, the shapes of the dendrites change, rewiring, in a sense, to code the new information.

Cell Body

The **cell body,** or **soma,** integrates all of the input signals from other neurons. Neurons either excite or inhibit each other, so the soma must determine which influence is greater. The soma also contains the nucleus of the neuron, where genetic instructions are stored and copied to make proteins.

Axons

The **axon** is a long arm extending from the soma. When excitation from inputs outweighs inhibition, a message is sent down the axon (called an **action potential**) that, in turn, reaches other neurons. The axon hillock is the transition of the soma to the axon, and it is this part that is critical to deciding whether or not a message is sent (see Figure 2.2). When the action potential reaches the end of the axon, chemical messengers called **neurotransmitters** are released.

Glial Cells

Glial cells are another type of cell in the nervous system. Unlike neurons, they do not carry information. Instead, they mainly serve a support role. There are many types of glial cells, each of which serves different functions. Some glial cells serve as scaffolding in the ner-

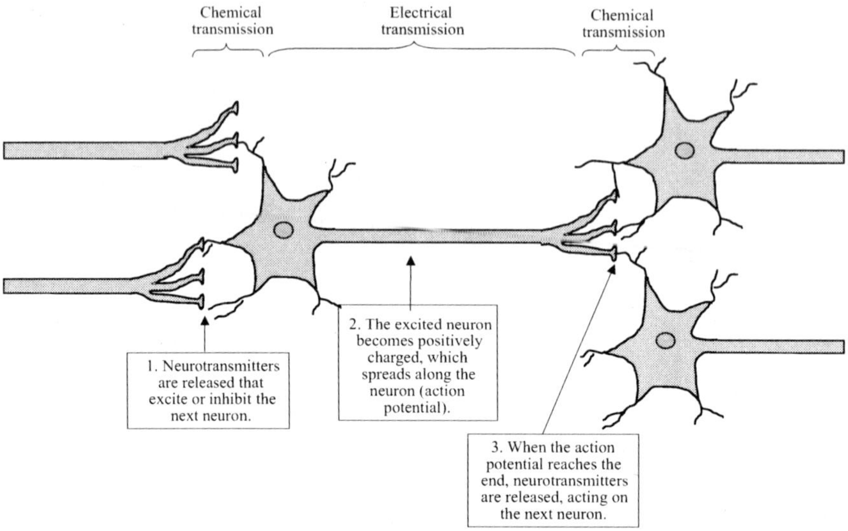

FIGURE 2.2. Neural communication.

vous system, providing structural support. They also help transport nutrients and wastes between neurons and the bloodstream. They also can store a small amount of energy and clean up the debris from dead neurons.

A certain type of glial cell wraps around axons and forms **myelin.** This serves as electrical insulation, making the action potentials faster and more efficient. Myelin is so important that neurons cannot perform their functions adequately without it. Diseases such as multiple sclerosis destroy myelin, causing a variety of neurological symptoms: The neurons remain intact, but the myelin is temporarily damaged and must regrow.

The Blood-Brain Barrier

The brain is a very sensitive organ, and many seemingly harmless things floating around the blood can cause it great harm. So the brain is protected by the **blood-brain barrier.** This barrier consists of extra-small pores in the blood vessels that supply the brain. In most of the body, the pores are large and most things can pass through. But in the brain, the pores are extremely small, so that only very small molecules (such as water) can pass through. The brain needs some large molecules to survive, such as amino acids and glucose. For these, the brain has transporters that only carry through the right molecules. However, in order to be psychoactive, drugs must be able to pass this barrier. By definition, all psychoactive drugs are highly fat soluble. This allows them to pass through the fatty membranes of cells and enter the brain. The more fat soluble a drug is, the more of it will get into the brain. For example, heroin and morphine are essentially the same drug. However, heroin is more fat soluble, which makes it many times more potent, ounce for ounce.

COMMUNICATION IN AND BETWEEN NEURONS

Communication among neurons is both an electrical and chemical process. Remember that neurons are connected end to end, in large networks. Dendrites receive the messages, somas measure excitation versus inhibition, and axons carry signals to the next neurons in the chain. Within the neuron, the messages are passed electrically (with charged ions), while between neurons the messages are passed chemically (with neurotransmitters).

Electrical Transmission

Neurons have a small voltage, which allows them to send electrical impulses. The neuron's voltage is created by an imbalance of charged particles called ions. When at rest, the neuron's charge is relatively more negative, which is called its **resting potential.** The neuron is quiet and not sending any signals. However, when excited by other neurons, certain ions enter the neuron and make it more positive. If the charge is sufficient, the neuron sends an electrical signal, the action potential. The action potential spreads along the axon, reaching the end, where it connects with other neurons. Afterward, the ions return to their original positions and the neuron returns to a negative, resting state. As mentioned previously, myelin makes this electrical transmission faster and more efficient.

Chemical Transmission

Chemical messages are passed between neurons by neurotransmitters. These are specially made chemicals that are stored inside the end of the axon. When an electrical signal comes down to the end of the axon, it triggers a release of neurotransmitters into the small space between the sending and receiving neuron called the **synapse** (see Figure 2.3). When a neurotransmitter is stored inside the sending neuron it is inactive, but when it is released into the synapse, it sends a signal by affecting the next neuron.

Neurotransmitter Receptors

Neurotransmitters influence neurons by attaching to **receptors** on the receiving neuron. When they find the right kind of receptor, it's like a key fitting into a lock—only the right key will fit. When the key (the neurotransmitter) fits into the lock (the receptor), it can either ex-

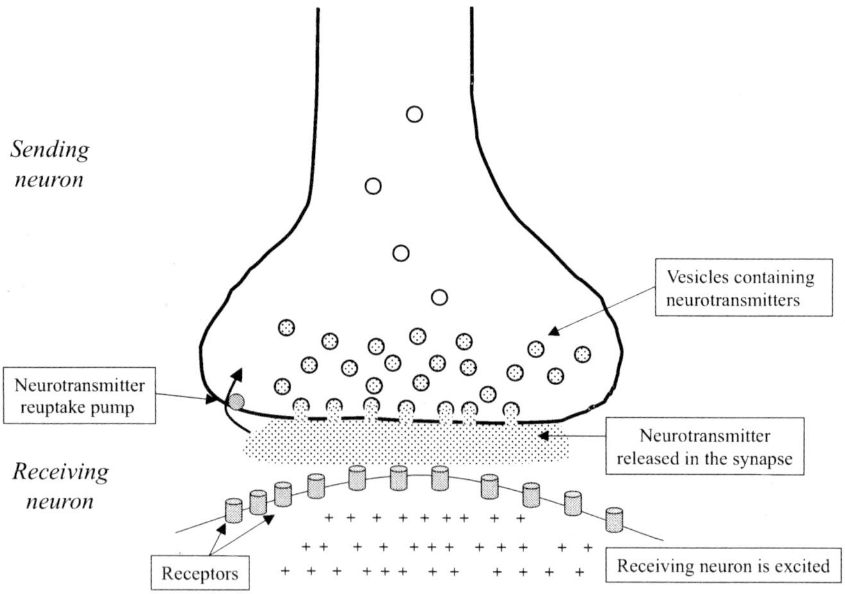

FIGURE 2.3. The synapse.

cite or inhibit the next neuron, depending on which neurotransmitter it is (some excite and some inhibit, while others can do both). Some receptors cause electrical changes that are needed for electrical transmission. Other receptors can only cause metabolic changes in the neuron (adjusting its metabolism or its sensitivity, for example).

The receptor is a major site where psychoactive drugs work. Many drugs (called **agonists**) imitate neurotransmitters. The receptor doesn't know whether a drug or a neurotransmitter is activating it—it only knows that it is being activated. For example, nicotine is an agonist drug at receptors for the neurotransmitter acetylcholine. Taking nicotine is tricking your neurons into thinking extra acetylcholine is present.

On the other hand, some drugs (called **antagonists**) attach to receptors but do not activate them. Instead they block them and keep them inactive. Antagonists, then, have the opposite effects of agonists. For example, heroin is an agonist drug. An overdose can stop a person from breathing and cause death. If the heroin user is given an antagonist drug in time, such as naltrexone, the overdose can be abruptly terminated and the person's life saved.

Neurotransmitters and Some Drugs That Interact with Them

Many neurotransmitters are found in the nervous system, and the list of new discoveries keeps growing. A few of the more common ones are discussed here. Many neurotransmitters are made from nutrients that come from our diet. This shows us how nutrition is necessary to healthy brain functioning.

Understanding neurotransmitters and the drugs that affect them is reciprocal: learning about one tells you about the other. Drugs that work on similar neurotransmitters have similar effects. Also, demonstrating their relationship to drugs can help people better understand neurotransmitters. For example, not many people know offhand what **dopamine** does, but when they know it's important for pleasure and addiction and is affected by drugs such as cocaine, it's easier to understand.

Acetylcholine

Acetylcholine (abbreviated ACh) comes from a vitamin called choline, which is in certain foods. It is released by nerves that control our muscles; any time you move your body, acetylcholine is being released. It also plays an important role in memory, and drugs that boost acetylcholine improve memory. Conversely, drugs that block it hinder memory. For example, drugs for Alzheimer's disease boost acetylcholine in the brain. Similarly, nicotine (from tobacco) is a drug that imitates acetylcholine, stimulating the brain and also improving memory.

Monoamines

The **monoamines** are a family of neurotransmitters including serotonin, dopamine, and norepinephrine. **Serotonin** is made from the amino acid L-tryptophan, while dopamine and **norepinephrine** come from the amino acid L-tyrosine, both of which derive from protein in the diet. These neurotransmitters affect diverse areas such as mood, appetite, and sex interest. For example, most antidepressant drugs boost these neurotransmitters.

Dopamine is especially involved in reward and reinforcement. It seems that every time we do something that is pleasurable, fun, or interesting, dopamine is released. This includes having sex, listening to enjoyable music, winning money at a slot machine, or eating a piece of chocolate. Dopamine ensures that we will repeat the behavior again. Some drugs (e.g., cocaine) hijack the dopamine system, activating it so much that they become addicting.

MAO (Monoamine oxidase) is an enzyme that breaks down all of the monoamine neurotransmitters.

Amino Acid Neurotransmitters

Certain neurotransmitters are actual amino acids, which also come from protein in the diet. One that excites the brain is called **glutamate.** It's found everywhere in the nervous system, so it plays a role in virtually all mental functions. Drugs such as **PCP** (known as phencyclidine, or angel dust) and **ketamine** (known as Special K) block glutamate. As a result, they cause hallucinations and strange dissociative symptoms, in which a user doesn't feel normal and experiences seem unreal.

GABA (gamma-aminobutyric acid) is a neurotransmitter with major inhibitory effects, which calm down neurons and keep them at rest. Accordingly, major sedative and anti-anxiety drugs (such as Valium, Xanax, alcohol, and phenobarbital) boost GABA's effects.

A careful balance must be maintained between neurotransmitters such as GABA and glutamate in the nervous system. With too much excitation, the brain becomes overactive and chaotic, resulting in a seizure. With too much inhibition, brain activity shuts down, stopping vital functions, resulting in coma or death (as in the case of alcohol overdose).

Neuropeptides

Neuropeptides are a class of neurotransmitters that are made of a chain of amino acids. One such neurotransmitter is called **substance P,** which sends pain signals to the brain. In contrast, **endorphin** is another neurotransmitter that has the opposite effect and inhibits pain. Thus, to stop pain signals from reaching the brain, you could either block substance P or boost endorphin. Another neuropeptide, called **oxytocin,** is becoming known as the "social bonding neurotransmitter." It is released during intimate contact and causes a warm glowing feeling that facilitates emotional bonding between people. For example, it is released when someone has an orgasm or is getting a massage. It is also released in mothers while breast-feeding, and encourages bonding between mother and baby.

Hormones

Several **hormones** are released by glands into the bloodstream. They regulate all kinds of body functions, but certain ones gain access to the brain and can alter the mind and behavior. Sex hormones, such as testosterone, estrogen, and progesterone, are known to do this. For example, it is well known that people who abuse anabolic steroids to build muscle also become more aggressive and irritable. Thyroid hormones affect brain metabolism; with too little people can seem sluggish and depressed, while too much can make them seem high-strung and nervous.

Gases

It may seem unlikely that a gas can be used as a neurotransmitter, but certain ones can, such as **nitric oxide** (NO) and even **carbon**

monoxide (CO). Ironically, both of these are toxic poisons in large amounts, but such small amounts occur in the brain that they are normally harmless. They also last only a few thousandths of a second before disappearing. Nitric oxide is not to be confused with nitrous oxide (laughing gas or N_2O). Nitrous oxide is an anesthetic used by some dentists, although the two closely related chemicals may interact. Nitric oxide also dilates blood vessels in the brain, tying blood flow to brain function: when a part of the brain becomes more active it makes more nitric oxide, which brings more blood to the area.

STRUCTURE OF THE NERVOUS SYSTEM

Beyond the fact that drugs are chemicals that alter the chemical workings of the brain, it also helps to know what effects they can have in different anatomical areas of the brain. For example, alcohol is known to cause a staggering walk and poor judgment, but these two effects are created in different parts of the brain. There is a division of labor in the brain—certain areas and systems carry out certain functions.

The central nervous system consists of the brain and spinal cord, while the peripheral nervous system consists of all the nerves running through the rest of the body. The nervous system is composed of white matter and gray matter. The gray matter is mostly somas of neurons, while the white matter is axons.

Spinal Cord

The **spinal cord** is the major route of input and output from the brain. Sensory nerves travel from the body to the spinal cord, carrying information up to the brain. Motor nerves descend from the brain down the spinal cord, and travel through the body, where they control muscles. Injuries that sever the spinal cord deprive a person of feeling and movement from that point downward.

The Autonomic Nervous System

The **autonomic nervous system** is a set of nerves that controls many automatic bodily functions, and governs essentially two states:

the **sympathetic** branch prepares us for physical action, while the **parasympathetic** branch prepares us for rest (see Table 2.1). For example, the sympathetic branch speeds up the heart rate while the parasympathetic branch slows it down. The sympathetic branch activates in response to threats, both physical attacks and psychological threats (e.g., anxiety).

The Brainstem

At the top of the spinal cord below the brain is the **brainstem.** This part of the brain carries out many primitive and vital functions, such as controlling breathing and heart rate. In the core of the brainstem is the **reticular formation,** which connects to higher areas in the brain and is responsible for maintaining a conscious state. When the reticular formation is inactive, a person lapses into unconsciousness, and when it is damaged the person remains unconscious in a coma or vegetative state. Many drugs have sedative effects by suppressing activity in the reticular formation. The **cerebellum** branches from the back of the brainstem and is involved in fine motor coordination. Damage to the cerebellum causes awkward, clumsy movements. Drugs such as alcohol suppress the cerebellum, causing the familiar drunken stagger. Most DWI (driving while intoxicated) tests used by police essentially measure how much the cerebellum is impaired by alcohol.

Structure of the Brain

The brain, or cerebrum, has two hemispheres, left and right, which are symmetrical. The hemispheres are interconnected, allowing con-

TABLE 2.1. The autonomic nervous system.

Function	Sympathetic	Parasympathetic
Heart rate	Speeds up	Slows down
Blood pressure	Raises	Lowers
Eye pupils	Dilates	Constricts
Lung bronchi	Dilates	Constricts
Digestion	Decreases	Increases
Salivation	Decreases	Increases

stant communication, and the largest connecting band is called the **corpus callosum.** The brain and spinal cord are covered by protective layers called the **meninges.** Infections of the meninges are known as meningitis.

The Cortex

The convoluted outer covering of the brain is called the **cortex** (Figure 2.4). It is divided into four lobes, which are symmetrical in both hemispheres: frontal, parietal, temporal, and occipital. The cortex is crucial for higher sensory and thought processes, and its wrinkled shape allows for maximum surface area inside the skull.

The **parietal lobe** is involved in body sensations and body image. The body you feel to be your own, even the hands you feel holding this book, are represented in the parietal lobe. The temporal lobe controls hearing, and the left side is involved in understanding language (either spoken or read). The **occipital lobe** is involved in vision.

The **frontal lobes** are more developed in humans than in other animals. They control voluntary movements, and damage to the frontal

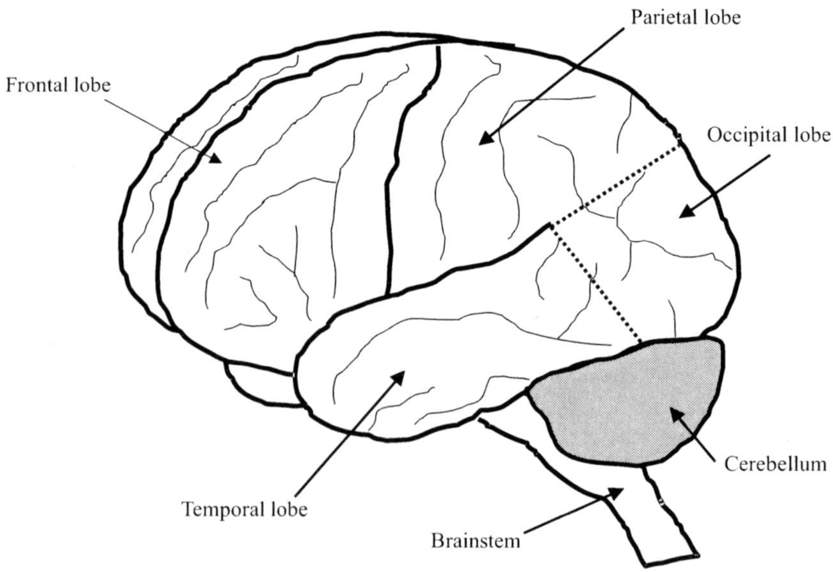

FIGURE 2.4. Side view of the brain.

lobe can cause weakness or paralysis. The left frontal lobe is involved in generation of language, whether spoken or written. Damage to that area produces an inability to formulate words and communicate. The most anterior portion of the frontal lobes (the prefrontal cortex) are involved in what are called executive functions. These are the most sophisticated forms of thought and include abilities such as abstract thinking, planning, organization, goal formation, social conduct, judgment, and decision making. For example, alcohol's impairment of one's judgment occurs through its effects on the prefrontal cortex. On the other hand, drugs such as Ritalin activate the prefrontal cortex, and are given to improve attention and impulsivity problems in kids with attention-deficit/hyperactivity disorder.

Ventricular System

Ventricles are fluid-filled chambers deep in the brain. The fluid produced in the ventricles is called **cerebrospinal fluid** (CSF). It circulates around the brain, supporting it and absorbing impact. Hydrocephalus is a condition in which the ventricles become clogged, and they swell outward, pressing against the brain tissue.

The Limbic System

The **limbic system** is a circuit of structures deep in the brain that mediate emotion. The **hippocampus** is a structure in the **temporal lobe** that is essential for memory. Drugs that enhance or hinder memory may act upon the hippocampus. The **amygdala** is another subcortical structure in the temporal lobe, which gives emotional meaning to things we perceive through our senses. Although it is involved in both positive and negative emotions, much research has been done on its role in anxiety and fear. Drugs such as Xanax suppress activity in the amygdala and alleviate anxiety.

Basal Ganglia

The **basal ganglia** are a set of structures that participate in cognitive and motor function. These structures degenerate in certain diseases such as Parkinson's disease, causing disorders of thinking and movement. Likewise, drugs that alleviate the symptoms of Parkinson's disease affect the basal ganglia.

Thalamus and Hypothalamus

The **thalamus** is an oval structure that serves as a sensory relay. All of the senses except smell connect with the thalamus first before activating their particular area of the brain. The thalamus also works with the reticular formation to keep the brain alert and awake.

The **hypothalamus** is found right underneath the thalamus, deep in the brain. It monitors many basic bodily activities and works to keep several functions within an optimal range. It regulates sleep, appetite, thirst, sex drive, and body temperature. It coordinates all of these functions with our emotions, behavior, and even hormones (via the pituitary gland). For example, aspirin lowers fevers caused by infection through actions in the hypothalamus. Also, some antidepressant drugs unfortunately reduce sex drive through their effect on the hypothalamus.

Chapter 3

How Drugs Work

Herbal medicines contain active drugs, in the true sense of the word. They are composed of chemicals that alter bodily processes. Since they are drugs, it makes sense to understand some basic aspects of the way drugs work.

Pharmacology is the study of drugs. It's divided into two main branches: pharmacokinetics (how the body affects drugs) and pharmacodynamics (how drugs affect the body).

HOW THE BODY AFFECTS DRUGS

Any drug ingested makes its way through the body. This includes how it gets into the body (absorption), how it moves around (distribution), how it is broken down **(biotransformation),** and how the body gets rid of it (elimination).

The importance of how a drug gets into the body and moves through the body cannot be underestimated. The intensity of a psychoactive drug depends on not only how much of a drug is taken (the dose) but how fast it reaches the brain. This is illustrated by the differences between chewing coca leaves, snorting powdered cocaine, and smoking crack. All three contain the same drug, cocaine, but it is delivered in different ways. As a result, chewing coca leaves causes milder stimulation, roughly equivalent to drinking strong espresso. Crack smoking, on the other hand, causes extremely intense stimulation and has become one of the worst addictions known. Some users have described the sensation as a "full-body orgasm." Other examples are morphine and heroin. These two are almost identical, with one minor modification. However, that modification makes heroin many times more potent than morphine, since it allows heroin to get into the brain faster and more thoroughly. As a result, ounce for ounce, heroin has far more intense effects.

Drug Absorption

Drug absorption is the process of a drug gaining access to the body. How a drug is taken is called the **route of administration** (Table 3.1). **Enteral** routes involve entry through the digestive tract. Swallowing a drug (oral administration) is the simplest and most common enteral route. **Sublingual** literally means under the tongue, but a drug is absorbed from any part of the mouth if it is not swallowed. An example would be chewing tobacco. Rectal administration means that a drug is inserted through the anus. This is usually reserved for a drug that irritates the stomach too much and causes nausea. An example is the curious practice of taking coffee enemas—the caffeine is absorbed just as well. When working in an epilepsy treatment unit, I saw a nurse with a "rectal Valium" injection kit. She explained that it was intended for patients having nonstop seizures. Since a person having a seizure can't swallow a pill, rectal administration is sometimes the easiest and next best way to administer medication. Enteral routes in general have the advantage of being cheap, easy, and generally safer. The disadvantage, however, is that they are less predictable: some of the drug may not be absorbed well, or it may be broken down in the stomach. It may also cause nausea.

Parenteral routes involve administering a drug to any part of the body other than the digestive tract. **Intravenous** injections skip the absorption process and introduce the drug directly into the bloodstream. This allows a physician to more accurately control how much drug a person is getting. Unfortunately, it also allows the possibility of massive doses of a drug, as in the case of heroin overdose. Vomiting is a safety mechanism that allows the body to be rid of a swallowed toxin, but this can't stop intravenous overdoses since they skip the digestive tract completely. **Intramuscular** means that a drug is injected into the muscle, from which it then enters the blood. **Subcutaneous** means a drug is injected right under the skin, from which it also slowly enters the blood. Heroin users call this method "skin popping." Intramuscular and subcutaneous routes are generally easier than intravenous injection, which requires correctly guiding the needle into a vein. Intraperitoneal means a drug is injected into the abdominal cavity, as are some chemotherapy drugs for cancer. Spinal injection means a drug is injected into the cerebrospinal fluid, near the spinal cord. This is often done to numb a woman from the waist down during childbirth.

Not all parenteral routes involve injection. Inhalation is probably the most common parenteral route: this includes smoking a drug. Inhaling (or smoking) a drug allows for very rapid and thorough absorption. Also, the blood from the lungs goes directly to the brain, so smoking a drug is really one step removed from injecting it directly into the brain (arguably much more difficult than smoking). Intranasal administration means that a drug is snorted and absorbed through the lining of the nasal sinuses. **Topical** (or **transdermal**) administration involves absorption from the skin (e.g., the nicotine patch to quit smoking).

Drug Distribution

Once a drug is absorbed, it goes into the bloodstream and spreads throughout the body. The better blood flow a person has, the faster the drug will be spread around. Since the brain, heart, liver, and kidney receive the most blood, they get most of the drug initially.

TABLE 3.1. Routes of administration.

Type	Subtype	Examples
Enteral	Oral	Swallowing a pill, drinking a tea
	Sublingual	Chewing tobacco
	Rectal	Enemas
Parenteral	Intravenous	Injecting heroin
	Intramuscular	Pain injection after surgery
	Subcutaneous	“Skin popping” heroin
	Intraperitoneal	Chemotherapy for cancer of the ovaries
	Spinal	Anesthesia for childbirth
	Inhalation	Smoking tobacco
	Topical/transdermal	The nicotine patch

Whether a drug reaches the brain (and thus is psychoactive) depends on whether it passes the blood-brain barrier. Since the brain is so sensitive, it must be protected from most substances in the blood by the blood-brain barrier. However, certain substances can get through. Anything that dissolves well in fat will pass through into the brain. By definition, most psychoactive drugs are fat soluble. Drugs that are fat soluble also pass well from a pregnant woman to the fetus.

Drug Biotransformation

Biotransformation involves the metabolism of the drug, or its chemical breakdown. This is accomplished mostly by enzymes in the liver whose goal is to make the drug more water soluble. Sometimes a drug is metabolized into another active drug. For example, not only is morphine psychoactive, but it breaks down into M6G, which also has psychological effects. The liver probably evolved in this role to remove the many toxic chemicals found in common plants we eat.

How fast a drug is metabolized is measured by its half-life. The half-life is how long it takes for the amount of drug in the blood to be reduced by half. Caffeine has a half-life of about five hours, so five hours after drinking a cup of coffee, half of the caffeine is still in the bloodstream. A steady state is achieved by taking a drug as fast as it's being metabolized, so that the amount in your body neither increases nor decreases over time.

Liver enzymes that break down a drug can increase when the amount of drug increases, as an attempt to compensate for the extra drug. This happens, for example, in chronic alcoholism. The liver works harder to process large amounts of alcohol. Eventually, this can result in damage to the liver, known as cirrhosis.

Excretion

Excretion is the process of eliminating a drug from the body. The liver makes drugs more water soluble, making it easier for the kidneys to filter them out of the blood. Once removed by the kidneys, drugs are eliminated in urine. Although the kidneys are the main route of elimination, there are others. Drugs are eliminated in virtually anything that comes out of a body: feces, hair, sweat, saliva, tears, and even air. The distinctive breath of a person who is drunk attests to the body's elimination of certain drugs by exhalation. Drug screening involves chemically detecting the presence of a drug or its metabolites eliminated from the body (as in urine or hair).

HOW DRUGS AFFECT THE BODY

For all drugs, it's important to understand the mechanism of action, or what it actually does in the body to create its effects. Drugs are understood in terms of their effects at a receptor, or a target to which it binds. Many different elements can serve as receptors, but most of them are some form of protein. This includes proteins that carry out chemical reactions (enzymes), transport chemicals in and out of cells (transporters), and allow the flow of charged particles in and out of cells (ion channels). By binding to a receptor on a protein, a drug can alter the way that protein functions. For example, a drug binding to an enzyme could stop it from breaking down a neurotransmitter, resulting in an accumulation of the neurotransmitter. A drug could also open an ion channel, exciting and activating a neuron. As mentioned in the previous chapter, a drug may act as an agonist at a receptor, imitating what a neurotransmitter would normally do. Or a drug may act as an antagonist at a receptor, attaching but keeping the receptor inactive (see Table 3.2).

Neurotransmitters and receptors may be thought of as keys and locks. Only the right drug or neurotransmitter will fit into the right receptor. For example, the neurotransmitter dopamine has its own specialized dopamine receptors but has no effect at serotonin receptors.

TABLE 3.2. Drug actions at receptors.

Action	Effect	Example
Agonist	Stimulates the receptor	Nicotine stimulates acetylcholine receptors
		Alcohol stimulates GABA receptors
Antagonist	Blocks the receptor	Naltrexone blocks heroin at opioid receptors
		Caffeine blocks adenosine receptors

Taking a drug repeatedly over time can change the sensitivity and even the number of the receptors. This is an example of neurons trying to adapt to the continued effects of a drug. Accordingly, if the drug is suddenly stopped, it takes time for the receptors to return to their original state.

Tolerance

Tolerance means that giving a drug repeatedly causes a progressively smaller effect. A person may find one cup of coffee very stimulating at first, but over time two or three cups may be needed to get the same effect. This happens partly because receptors in the brain are adapting to the presence of the drug, so higher doses are needed to maintain the same level of effect. Another form of tolerance occurs when the liver increases its activity to break down the drug faster. For example, a heavy drinker can break down alcohol much faster than a nondrinker. In most cases, however, tolerance can be reversed over time if the drug is no longer taken.

Dependence

Dependence means that a drug is needed to keep functioning normally, and discontinuing the drug causes a **withdrawal syndrome.** Some people distinguish between physical dependence (characterized by physical withdrawal symptoms, such as a headache) and psychological dependence (needing the drug to function mentally), but, in reality, both of these are closely related. For example, caffeine withdrawal involves physical symptoms such as headache and psychological symptoms such as lethargy. Both of these discourage a person from ceasing to use it.

Interactions Between Drugs

Whenever two or more similar drugs are given in combination, an interaction can occur. For example, giving two antianxiety drugs together, such as Xanax and Valium, can produce additive effects. The effect of both is like adding together the individual effects. However, in some cases, **synergy** can occur, as with Xanax and alcohol. They increase each other's effects, but instead of addition it's more like multiplication. Drugs such as Xanax can be very safe by themselves but dangerous when combined with a sedative such as alcohol.

RELATIONSHIPS BETWEEN DRUG DOSES AND EFFECTS

A final branch of pharmacology examines the relationship between the dose of a drug given and the intensity of effects it produces. To show this graphically, we use a **dose-response curve** (see Figure 3.1). This type of graph shows what we all know intuitively: the more you take, the stronger the effect. Taking one aspirin relieves a little pain, while taking two aspirin relieves more. The S shape of the curve also confirms what we know about ceiling effects: no matter how much you take, the drug can only do so much. For example, aspirin reduces pain, but only to a point. Increasing the drug dose will not increase the effect indefinitely.

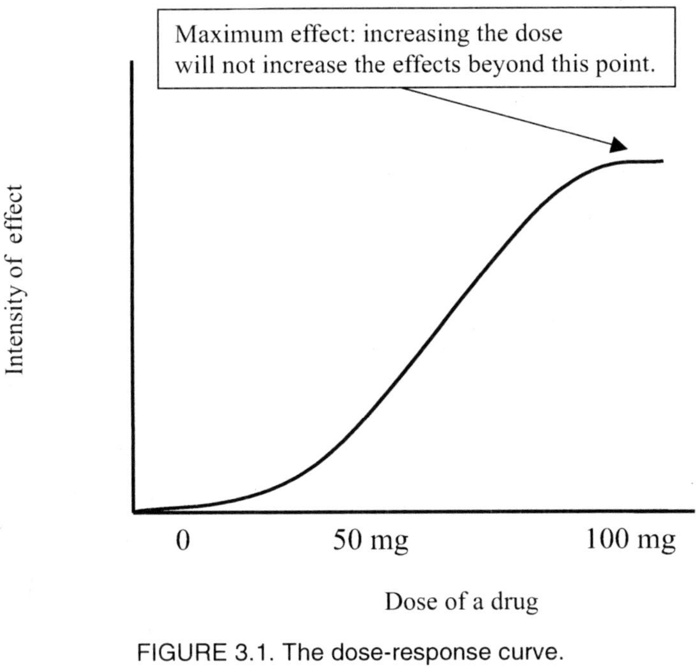

FIGURE 3.1. The dose-response curve.

However, in certain cases, instead of an S shape, the curve looks like an upside-down U (see Figure 3.2). Sometimes medium doses are better than small or large doses. This may sound strange at first, but consider the example of coffee and concentration ability. Small amounts of coffee stimulate the brain and can improve your ability to concentrate, especially if you are tired. Two cups are stronger than one, and your alertness and ability to concentrate increase. However, you can imagine what would happen if you kept drinking more and more cups. Eventually you would feel completely wired, jittery, and even anxious. Then your concentration ability would decrease, which is the downslope of the inverted U.

Drugs can differ in terms of **potency** and **efficacy.** Potency refers to the strength of a drug, or how much you need to get a certain level of effect. For example, in Figure 3.3, graph A, Drug X is more potent than Drug Y because it accomplishes the same effect at a lower dose. With a high enough dose, however, they can both have the same effect. Efficacy, on the other hand, refers to the maximum effect a drug can have. In Figure 3.3, graph B, Drug P has more efficacy than Drug Q because at the same dose, P has a more intense effect. No matter how much of Drug Q you give, it will never reach the maximum that P can.

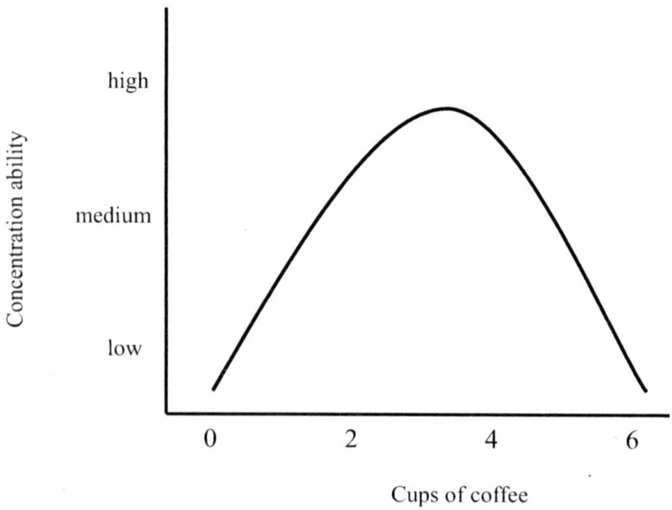

FIGURE 3.2. Example of an inverted U dose-response curve.

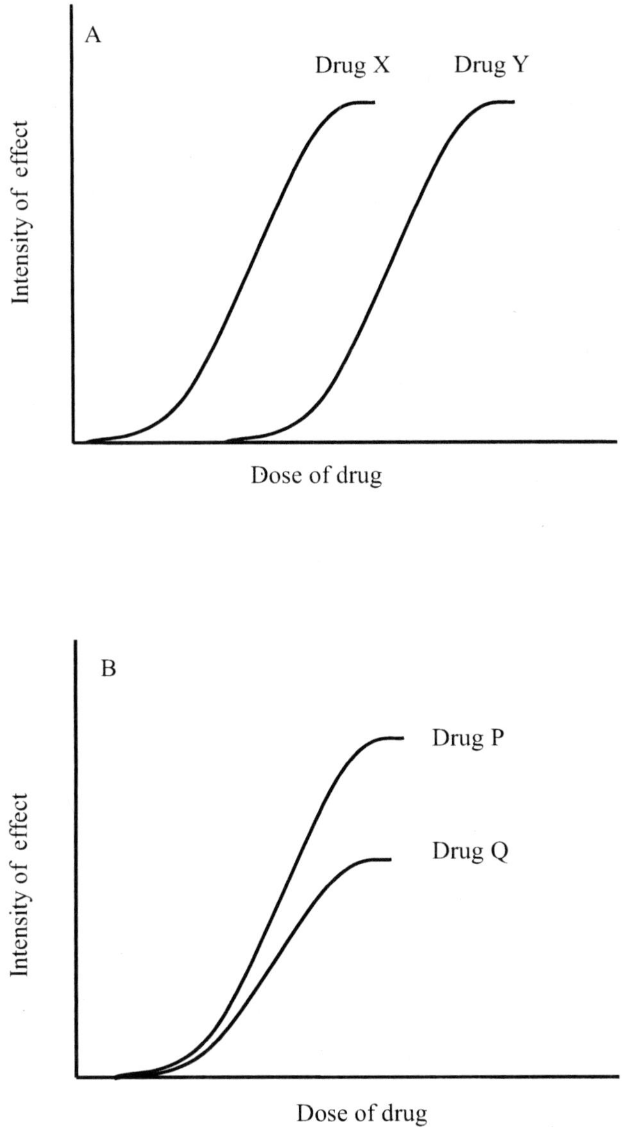

FIGURE 3.3. Differences in potency versus efficacy. A: Drugs X and Y will both create the same maximum effect in sufficient doses. However, Drug X has greater potency since it accomplishes the same thing at a lower dose. B: Drug P has greater efficacy than Drug Q. No matter how high the amount of Drug Q, the ceiling for it is lower.

Drug Safety Studies

In order to look at the safety of a drug, two dose-response curves are compared: one for the **effective dose** (the amount needed to create the desired effect) and one for the **lethal dose** (the amount that will kill). The farther apart these two curves look, the safer the drug. For example, if a drug is effective at 10 mg and lethal at 10,000 mg, then it is an incredibly safe drug. It would be very difficult to accidentally overdose on the drug. However, if another drug is effective at 10 mg but lethal at 20 mg, then it is easy to see how taking just a little bit too much can be dangerous.

Chapter 4

Stimulant Herbs

There's something fundamentally appealing about **stimulant** plants. The ability to survive and flourish depends on one's ability to mentally focus, and sometimes remain focused for long periods of time. Stimulant drugs can assist us in this regard, increasing alertness, reducing fatigue, and allowing us to sustain effort. In some cases, they elevate mood and cause euphoria.

Stimulants are among the earliest psychoactive plants used by humans. They are also the most widely used psychoactive drugs, so common that many people easily forget they are drugs. Billions of people use tobacco, coffee, tea, or some other form of stimulant on a daily basis. Depending on the plant, culture, time period, and context of use, stimulants have been regarded as indispensable items or a deadly scourge. For example, coca was considered a divine drug by the Incas, and to use it was a privilege. Today, the availability and potency of purified cocaine make it a socioeconomic hazard.

Stimulant plants are organized in this chapter by function (see Box 4.1). Similar chemical mechanisms in the brain means that they have similar psychological and behavioral effects. For example, coffee, tea, cocoa, maté, guarana, and cola all work through common mechanisms. Tobacco, areca, and lobelia form a second cluster of stimulants. Ephedra, coca, and khat form the third group.

PHARMACEUTICAL STIMULANTS

Many synthetic stimulant drugs have been invented for medical and other uses. Amphetamine was created in 1887, but its medical uses were not investigated until the 1920s. It was first sold in 1932 to treat asthma and nasal congestion. It was later used for appetite re-

BOX 4.1. Classification of Stimulant Herbs

I. Caffeine stimulants (block adenosine)
   A. Coffee
   B. Tea
   C. Cocoa
   D. Guarana
   E. Maté
   F. Cola (or kola)

II. Tobacco-like stimulants (activate acetylcholine systems)
   A. Tobacco
   B. Lobelia
   C. Areca

III. Amphetamine-like stimulants (activate norephinephrine and/or dopamine systems)
   A. Ephedra
   B. Coca
   C. Khat

duction and weight loss. However, addiction potential, tolerance, and side effects limit its use. Further, once the drug is stopped there is a tendency to regain the lost weight.

A number of related drugs were developed for attention-deficit/hyperactivity disorder (ADHD) after amphetamine: methylphenidate (Ritalin), pemoline (Cylert), and atomoxetine (Strattera). Adderall is a mix of different amphetamine salts. ADHD is one of the most common clinical conditions for which stimulants are prescribed. It is characterized by inability to sustain attention, excessive physical activity or talkativeness, and impulsivity. It may seem paradoxical to give a stimulant to an overactive child or adult, but the condition is associated with underactivity in the frontal lobe systems of the brain, which are involved in self-control and concentration. Thus, the right dose of a stimulant may bring brain activity up to normal levels.

Ecstasy (methylenedioxymethamphetamine or MDMA) and several other similar drugs were developed from amphetamine. These drugs are called entactogens, because they create feelings of warmth and empathy between people. Unfortunately, they also are highly neurotoxic to serotonin neurons. The long-term effects of Ecstasy in humans are not yet known. Regular users of Ecstasy are essentially a large group of guinea pigs.

Despite the existence of multiple synthetic stimulants, plants or plant-derived drugs probably still account for the majority of human consumption of stimulant drugs.

CAFFEINE STIMULANTS

The **caffeine** stimulants are a diverse group of plants that contain caffeine and similar stimulant drugs. Despite the fact that they are different plants and come from different parts of the world, they all have similar effects because of the common drugs they contain: caffeine, theophylline, and theobromine. These are probably the most commonly used stimulants in the world, and maybe even the most commonly consumed of all drugs. The plants that contain caffeine and related stimulants are coffee, tea, chocolate, guarana, maté, and kola. The plants are discussed individually first, and then the effects of caffeine are covered together.

Coffee

The coffee plant *(Coffea arabica)* grows dense clusters of flowers and its famous oval bean (see Figure 4.1). The bean is green initially, changing to yellow and red with maturity, and turning brown when roasted. The plant is native to Ethiopia but is cultivated in many tropical regions of the world. It is believed that the name *coffee* comes from the Turkish word *gahveh* or Arabic *gahuah*. When it was introduced to Europe in the seventeenth century, it was met with public resistance before it eventually gained acceptance.

Active Chemicals in Coffee

Coffee contains caffeine, as well as its close chemical relatives theobromine and theophylline. The amount of caffeine varies with the preparation: drip-brewed coffee usually has 60 to 180 mg per five-ounce cup, while instant coffee may have 30 to 120 mg. "Decaffeinated" coffee is not completely free of caffeine, having 5 mg or less. This can add up after drinking a few cups, especially if one is not expecting to get any caffeine. For comparison, over-the-counter caf-

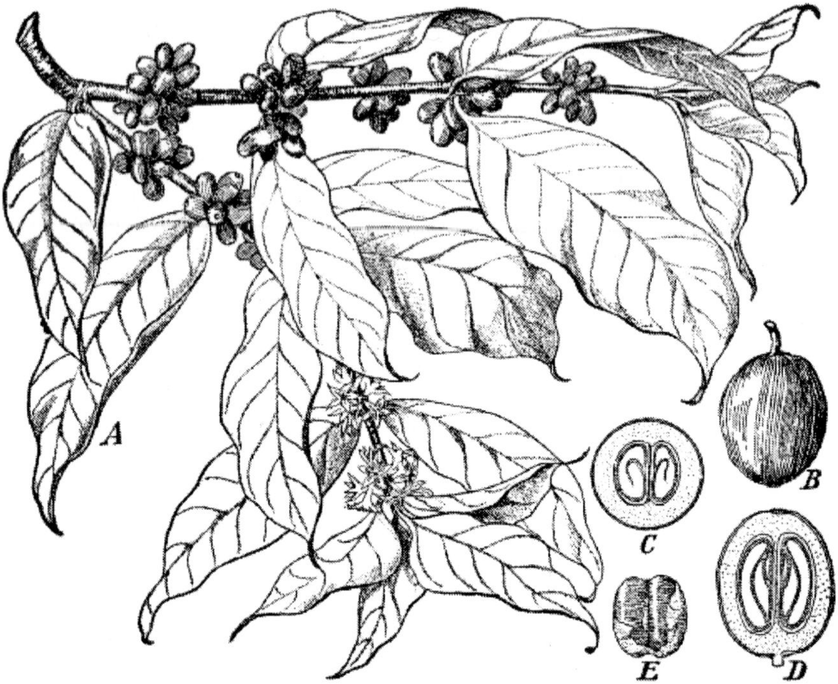

FIGURE 4.1. Coffee *(Coffea arabica).* Reprinted from Culbreth DMR (1927). *Materia Medica and Pharmacognosy,* Seventh Edition. Philadelphia: Lea & Febiger.

feine pills (No-Doz, Vivarin) have about 100 mg per tablet, and caffeinated soft drinks have 40 to 60 mg of caffeine per twelve-ounce serving. Caffeine pills may have the same amount of caffeine as an average cup of coffee, but because they are in a purified, concentrated form, the stimulant effects can come on fast and strong.

Tea

Both black and green tea come from the same plant, *Camellia sinensis*. Tea is made from soaking the dried leaves in hot water. Green tea is made by simply drying the leaves after they are picked, while for black tea the leaves are picked and fermented before drying. The plant was originally cultivated in China and now grows in many parts of the world. It is one of the most commonly consumed drinks worldwide.

Active Chemicals in Tea

Similar to coffee, the major active drugs in tea are caffeine, theobromine, and theophylline. A cup of brewed tea has approximately 2 to 100 mg of caffeine per five-ounce cup, while instant tea tends to have less (20 to 50 mg per cup). Tea also contains a significant amount of antioxidant chemicals, which are being researched for their health effects.

Cocoa and Chocolate

Of all the stimulants, cocoa may be said to be the tastiest. Pure cocoa powder comes from the plant *Theobroma cacao* (see Figure 4.2). The word *theobroma* comes from Greek, meaning "food of the gods," and *cacao* was the Aztec name for the plant.

The seeds of the cacao plant are removed from their shells, then fermented and ground into a powder. Pure cocoa powder is bitter. Cocoa is best known in sweetened chocolate; chocolate derives from the Aztec word *chocolatl.* The Spanish conquistador Cortés brought co-

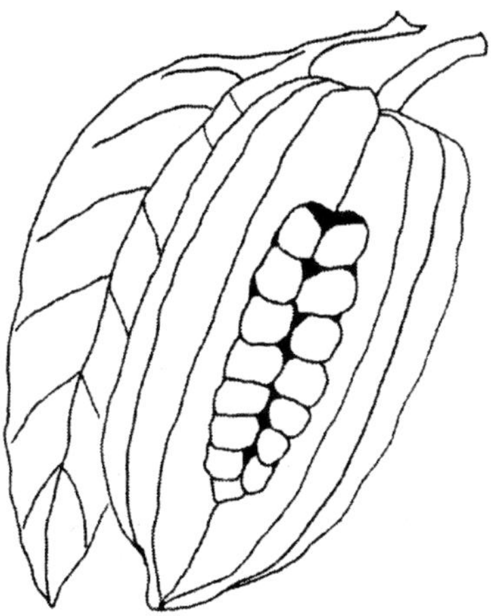

FIGURE 4.2. Cacao bean *(Theobroma cacao).* Copyright 2004 by Marcello Spinella.

coa back to Europe in 1528, but it didn't become popular until the seventeenth century. At the time it was consumed as a bitter drink, mixed only with hot water but no sugar. The Swiss changed the history of candy by adding sugar and milk solids to make milk chocolate. Milk chocolate was first sold in 1876 under the brand name Nestlé.

Active Chemicals in Cocoa

Chocolate has a significant amount of theobromine, but it also contains caffeine. An average pack of instant hot cocoa mix has about 60 mg of theobromine and 4 mg of caffeine per serving. Chocolate generally has less caffeine than coffee; dark chocolate can have significant amounts. Milk chocolate has up to 15 mg of caffeine per ounce, and dark chocolate has up to 35 mg per ounce. So, three ounces of dark chocolate has the caffeine equivalent of an average cup of coffee.

Guarana

Guarana *(Paullinia cupana)* is a climbing shrub that grows in the Amazon basin. It produces a purple-brown seed about the size of a hazelnut. Indians of the Amazon prepare the seeds by roasting them and grinding them into a paste. They typically use it for stimulation in daily tasks such as hunting. In more recent years, guarana has become available and more popular in the West.

Active Chemicals in Guarana

Since it is relatively less known than tea or coffee, many people are unaware that guarana contains caffeine. It also has theobromine and theophylline in smaller amounts. One gram of seed typically contains 35 to 55 mg of caffeine.

Maté

Maté, a popular herb in South America, is also known as yerba maté or Paraguay tea. The dried leaf and stem of *Ilex paraguariensis* are used to make a tea. However, tea bags are not used; instead, hot water is poured directly over dried leaves in a cup. A metal straw with a strainer at the end, called a *bombilla,* is used to drink the tea, keeping out the leaves.

Active Chemicals in Maté

Maté contains both caffeine and theobromine. Like guarana, since it is less commonly known in North America, many people are unaware that it contains caffeine.

Kola

Kola (also spelled cola) is a nut that comes from *Cola nitida* and *C. acuminata*. The tree is native to West Africa, but it is also cultivated in other similar climates across the world. The seeds (or nuts) of the kola tree are taken from the pods and their outer coat is removed. They must either be chewed fresh or dried as they otherwise spoil quickly.

In Africa, the nuts are used for their stimulant effects. They were traditionally valued and exchanged as ceremonial gifts. However, kola has been more widely consumed in the form of Coca-Cola. As the name suggests, the original beverage contained extracts from the coca shrub *(Erythroxylon coca)* and the kola nut, making a small but potent combination of caffeine and cocaine. The drink, which was initially green, was first sold in 1886. Cocaine was later eliminated from the drink around the turn of the twentieth century. Decocainized coca extract is still used by the Coca-Cola company for flavoring.

Do not confuse kola with the similar-sounding herb gotu kola *(Centella asiatica).* This is an entirely different herb that does not contain caffeine.

Active Chemicals in Kola

Kola nuts contain caffeine and smaller amounts of theophylline and theobromine. The fresh nut contains a chemical (called kolacatechin) that splits into caffeine and theobromine in the drying process. The drying process also gives kola its characteristic red-brown color, which is now imitated by caramel coloring in cola soft drinks.

Drug Issues for Caffeine

Caffeine is rapidly absorbed from the digestive tract when taken orally. Food generally slows this process, so taking caffeine on an empty stomach has more rapid effects. Caffeine spreads to all parts of the body and can cross into fetuses and breast milk. The half-life of caffeine is three to seven hours, but it may be longer in women using oral contraceptives or in late-stage pregnancy.

How Caffeine Works

Caffeine works by blocking a neurotransmitter called adenosine in the brain, which gives a negative feedback signal to the neurons. This means it slows down neurons and has a sedating effect. However, since caffeine blocks adenosine, it reduces the sedative effects of adenosine and has a stimulant effect.

General Effects

Smooth muscle. Caffeine and its relatives relax smooth muscle, the type found in many internal organs and blood vessels. In the lungs this opens the air passages, so the caffeine relative theophylline has been used as a treatment for asthma. Although caffeine relaxes arteries in the heart, it constricts blood vessels in the brain.

Diuretic effects. Caffeine increases the production of urine. It also increases blood flow to the kidneys and the speed at which they filter blood. Long car trips fueled by coffee are also frequently interrupted by rest room stops.

Cardiovascular effects. Caffeine and its relatives have complex effects on blood circulation. In addition to directly affecting blood vessels and the heart, they also alter blood flow through actions in the brain and hormonal glands. Overall, caffeine generally increases blood pressure. High doses of caffeine can cause rapid heartbeat and abnormal heart rhythms.

Hormonal. High doses of caffeine produce a stresslike hormonal reaction. This entails a release of hormones from the adrenal gland, such as cortisol. These effects occur when one takes approximately 500 mg, the equivalent of five average-strength cups of coffee. Caffeine is removed from the body at a slower rate in women during the late luteal (premenstrual) phase. So caffeine may have an extra kick at this point of a woman's cycle.

Muscular effects. Caffeine increases the work capacity of skeletal muscle, the type we use for moving our bodies. Whether or not caffeine actually improves exercise is debated. There doesn't seem to be any benefit for high-intensity, short-term exercise (such as sprinting), but it may help in prolonged, moderate-intensity exercise (such as jogging).

Behavioral effects. Caffeine improves motor performance, affecting abilities such as manual dexterity. The benefits are most noticeable when a person is fatigued.

Although it may seem counterintuitive, caffeine does not increase disruptive behavior in preschool children. Large-scale studies show that when objectively measured, regular doses of caffeine do not have any negative cognitive or behavioral effects in groups of children.

This is not to say that caffeine has no effects in children. In fact, the individual effect may vary from child to child (as in adults). An important factor is how much the child normally consumes: children who normally consume high levels of caffeine (500 mg/day or more) become more anxious when without caffeine. When given caffeine (5 mg per kilogram of body weight), those who rarely use caffeine become more emotional, inattentive, and restless. Their high-caffeine-consuming counterparts, however, do not show a difference. So a child's reaction to caffeine certainly differs according to how much he or she regularly takes. It may also vary by some inherent differences among children. When low doses of caffeine are added to Ritalin in children with ADHD, it causes greater improvement than Ritalin alone.

Although some people take caffeine to sober up after drinking alcohol, this probably does not work. Caffeine may make people more awake and less likely to fall asleep. However, the impairments in thinking, coordination, and reaction caused by alcohol are not reduced.

Sleep and arousal. As anyone who has taken caffeine at night can attest, it potently increases a person's level of arousal and wakefulness. In practical terms, this means increased alertness and taking longer to fall asleep. Many people claim that drinking a caffeine beverage, such as coffee, has no effect on them, and that they can fall asleep soon after drinking it. There are two things to consider here. The first concerns tolerance: people who drink coffee regularly are not as strongly affected as light drinkers or nondrinkers. More important, even though they fall asleep, caffeine still affects the depth of sleep and how restful it is. Although they may sleep a full eight hours, caffeine makes them spend more time in the light, superficial stages of sleep.

Emotional. Caffeine has mixed effects on emotion. Smaller amounts (around 250 mg) make a person feel more energetic and positive. However, as the dose gets higher (e.g., 500 mg), so does the likelihood of anxiety, tension, and nervousness. People with anxiety or panic disorders are even more sensitive to the anxiety-causing effects of caffeine.

Brain activity. As one would expect, caffeine increases brain activity. This is seen on the **electroencephalogram** (EEG), which shows faster activity in the brain after one consumes caffeine. Quitting tobacco causes slowing of activity on the EEG, but caffeine can bring it back to normal. Conversely, quitting caffeine can slow down brain activity.

Caffeine increases the amount of energy used by the brain, which is also consistent with its stimulant effects. Drugs that increase energy use by the brain tend to increase blood flow to the brain: more blood to deliver more energy. However, caffeine paradoxically decreases blood flow to the brain, because it is a vasoconstrictor. Caffeine acts directly on blood vessels to constrict them.

Cognitive effects. As expected from a stimulant, caffeine improves reaction time and attention, problem solving, and memory. Even though these effects may diminish to some degree with chronic caffeine use (i.e., tolerance), improvements in attention are sustained. On the other hand, withdrawal from caffeine impairs some mental abilities. Caffeine seems to be the essential ingredient for mental alertness, since black tea and coffee have similar cognitive effects. The results are greater in females, but this varies with the menstrual cycle. For unknown reasons, combining caffeine with aspirin boosts the actions of caffeine. How this happens is uncertain, since aspirin is not a stimulant by itself.

Research shows that caffeine really does reduce the mental impairments caused by lack of sleep. It also reduces fatigue, making concentration easier.

Addiction and Dependence

The patterns of caffeine use in some individuals may be classified as abuse or addiction. Chronic use causes tolerance, and discontinuation causes a withdrawal syndrome. The intensity of the withdrawal syndrome varies with the amount and duration of caffeine use. It consists of anxiety, lowered mood, fatigue, and headache. This is reflected by altered adenosine receptors in the brain, where caffeine works. In order to reduce or avoid the withdrawal syndrome, the dose must be slowly reduced, allowing the receptors in the brain time to gradually return to normal.

Side Effects

Normal doses of caffeine do not cause serious toxicity. Fatal poisoning is rare, and requires enormous doses of 5 to 10 g (over fifty cups of coffee). However, negative side effects still occur at doses far below the fatal dose. At doses of 1 g (about ten cups of coffee), people experience overstimulation, insomnia, rapid heartbeat, and vomiting, which may progress to delirium and seizures.

A small increased risk of cardiovascular disease is associated with long-term use of caffeine. Coffee seems to contain a noncaffeine chemical that raises cholesterol levels. Fortunately, it is reduced by coffee filters.

Caffeine is probably the most commonly used drug among pregnant women, but its safety during pregnancy is still debated. Caffeine use during pregnancy does not seem to alter children's IQ and attention ability at seven years of age. However, it may be related to reduced fertility, miscarriage, growth retardation in utero, and sudden infant death syndrome.

A high dose of caffeine can induce a seizure in anyone, but such doses are rarely if ever taken. However, caffeine can worsen seizures in people who are prone to them. For example, caffeine prolongs the seizures that are induced by electroshock therapy for depression. Thus, caffeine is best avoided by people with epilepsy.

TOBACCO AND TOBACCO-LIKE STIMULANTS

Two plants are tobacco-like stimulants: **lobelia** and **areca.** They have effects similar to tobacco because they stimulate the same neurotransmitter system in the brain: acetylcholine. Although they activate the same system in slightly different ways, their psychological effects are still very much alike.

Tobacco

The tobacco plant *(Nicotiana tabacum)* is a flowering annual or biennial plant (see Figure 4.3). The leaves range from ovate to narrow and tapering, and grow up to 50 cm in length. The tobacco plant is native to tropical regions of America, but is now grown in many parts of the world, including China, Turkey, Greece, Holland, France, Germany, and subtropical regions. The leaves of the plant are dried for smoking or chewing purposes.

Native Americans used the tobacco plant long before the arrival of Europeans. Its name derives from *tabacum,* the Indian name for the pipe they used to smoke it. **Nicotine,** the main active drug in tobacco, was named after Jean Nicot de Villemain, a French diplomat who popularized tobacco in Europe in the late 1500s. Less commonly known is that tobacco grew wild in the interior of Australia and was used by inhabitants there before Europeans arrived.

Active Chemicals in Tobacco

The main psychoactive drug in tobacco is nicotine. The plant contains similar drugs including nornicotine and cotinine, among others. However, nicotine alone is sufficient to create the psychological effects commonly associated with tobacco.

How Tobacco Works

Nicotine creates stimulant effects by activating the acetylcholine system in the brain. In fact, the receptor it attaches to is called the nicotinic receptor, because it was discovered with nicotine. When nico-

FIGURE 4.3. Tobacco *(Nicotiana tabacum).* Reprinted from Culbreth DMR (1927). *Materia Medica and Pharmacognosy,* Seventh Edition. Philadelphia: Lea & Febiger.

tine attaches to this receptor, it excites neurons, increasing their activity and producing mental stimulation. So, in a sense, taking nicotine is tricking the brain into acting as if it had extra acetylcholine. However, nicotine attaches so tightly that it desensitizes the receptor. So initially nicotine activates it and then makes it inactive over time. Nicotinic receptors are found in many areas of the brain and are especially involved in functions such as attention and memory.

Although nicotine directly excites the nicotinic receptor, it also has indirect effects on many other systems. There are too many to discuss here, but one deserves mention. Nicotine, like many other drugs with addiction potential, increases the neurotransmitter dopamine. This is probably why nicotine can be so hard to give up.

Drug Issues of Tobacco

As mentioned before, the intensity of a psychoactive drug is determined by how much is taken and how fast. When tobacco is smoked, nicotine very quickly passes into the blood, creating an intense effect. When it is chewed, the effect is less intense but more prolonged. It absorbs through the gums and mouth (e.g., chewing tobacco, nicotine gum) and through the skin (e.g., the nicotine patch). However, nicotine does not absorb well from the stomach, which is why there is no nicotine pill to help people quit smoking.

An average cigarette contains about 12 mg of nicotine, and the smoker actually gets about 1 mg. People seem to adjust their rate of smoking to keep a steady level of nicotine in the blood: when people use low-nicotine cigarettes, they smoke more of them to make up for it.

Effects of Nicotine

Cardiovascular. Nicotine increases heart rate and blood pressure by activating the autonomic nervous system.

Digestive tract. Tobacco increases bowel activity, moving food through at a greater rate. First-time users of tobacco often experience nausea, vomiting, and occasionally diarrhea.

Hormonal. Tobacco smoking increases several hormones, including stress hormones such as cortisol. Quitting smoking, conversely, decreases stress hormones.

Muscular effects. Acetylcholine and nicotinic receptors are used by nerves that control our muscles. Every time we make voluntary movements, our nerves are releasing acetylcholine to make muscles contract. Although nicotine from tobacco first stimulates this system, it quickly makes it less sensitive. The result is that nicotine has muscle relaxant effects.

Behavioral and emotional effects. Nicotine has many behavioral and emotional effects. It suppresses appetite and decreases weight gain. The flip side of this is the well-known weight gain that can occur after quitting smoking. Nicotine reduces anxiety in humans, more so in females than males. This has been measured by the brain electrical activity (EEG) of people watching a stressful movie. Again, quitting smoking has the opposite effect of increasing anxiety and reducing happy mood.

Brain activity. As would be expected of a stimulant, tobacco activates electrical patterns in the brain. It reduces slow-wave patterns and increases the fast-wave patterns associated with thinking and wakefulness. Quitting smoking has the opposite effect, at least temporarily, slowing down brain electrical activity. Also as expected, nicotine gum has the same effects as nicotine cigarettes. In contrast, cigarettes with the nicotine removed do not activate brain electrical patterns.

Nicotine increases blood flow and energy usage by the brain, also typical of a stimulant drug. It activates the limbic system in a way consistent with improving mood and making people feel good.

Dependence and Addiction

Nicotine is clearly addicting in every sense of the word, in both animals and humans. When regular smokers quit, they experience a withdrawal syndrome consisting of drug craving, depression, anxiety, irritability, problems with concentration and memory, and poor-quality sleep. The severity of this syndrome depends on how much nicotine has been used and for how long. Not surprisingly, these symptoms make quitting difficult and result in many failed attempts to quit. Most of these symptoms disappear within a month after quitting, but weight gain and increased appetite can last for six months. Some smokers report that the craving may last for years or indefinitely.

Depression is common among smokers. Smokers with depression have more intense withdrawal symptoms, less success at quitting, and a greater chance of relapse. It has been suggested that smoking is actually a way to self-medicate for the low mood and thinking difficulties that are characteristic of depression.

The addictive effects of tobacco occur through both positive and negative reinforcement. Positive reinforcement happens whenever doing something makes you feel good. Nicotine has stimulant effects and lifts mood, which positively reinforces smoking. Negative reinforcement, on the other hand, happens when doing something alleviates unpleasantness or suffering. So using tobacco to reduce anxiety or relieve withdrawal symptoms is negative reinforcement. (Don't confuse negative reinforcement with punishment. Punishment occurs when an action is followed by adverse consequences, making it less likely that the behavior will be repeated. Negative reinforcement takes place when an action results in the removal of an aversive experience.) These reinforcing effects of nicotine are likely a result of its effects on the neurotransmitter dopamine, which plays an important role in reinforcement.

Addiction to nicotine occurs over time; the longer a person smokes, the more likely he or she is to become addicted. Smoking more also makes it harder to quit: the number of cigarettes smoked per day is related to the number of unsuccessful attempts to quit.

Drugs to Help People Quit Using Tobacco

Nicotine replacement. Most treatments for tobacco addiction focus on replacing the nicotine. This allows a person to break the habit of smoking without tackling the nicotine problem at the same time. Then the nicotine can be slowly reduced instead of having a full-blown withdrawal syndrome. The most common forms of nicotine replacement are a patch and gum. The main problem with these is that they provide a slow absorption of nicotine, while cigarettes have a rapid effect. The most recent solution to this problem is a nicotine inhaler, so a user can take nicotine in through the lungs without smoke. Still, the patch and gum reduce the severity of withdrawal symptoms.

Other drugs for tobacco addiction. A few other drugs have been tried to ease tobacco withdrawal symptoms. Clonidine, a blood pressure medication, has shown mixed or small effects. BuSpar is an antianxiety drug that has been tried since anxiety is a prominent feature of tobacco withdrawal. However, results have been mixed and more research is needed.

The best non-nicotine drug for treating tobacco addiction seems to be Zyban. Zyban is the trade name for the drug bupropion, which is also sold as an antidepressant under the trade name Wellbutrin. It elevates brain levels of the neurotransmitter dopamine. It was reasoned that if nicotine addiction is related to increased dopamine activity, and withdrawal causes low dopamine activity, then boosting dopamine with Zyban could help people quit. This seems to be true: controlled studies show that Zyban helps people quit using tobacco. Further, it can be used in combination with the patch or gum, and is more effective than the patch alone.

Toxicity

High doses of nicotine are very toxic. The lethal dose in adult humans is approximately 60 mg. Although this dose is rarely reached, it has happened. One case involved a boy taking a bet that he could smoke 100 cigarettes in a row, which resulted in his death. Acute nicotine overdose causes headache, dizziness, salivation, cold sweats, abdominal pain, nausea, vomiting, and diarrhea. Confusion and seizures may also occur. Blood pressure drops and the heartbeat becomes weak and irregular. Death usually results from suppression of breathing.

Death from short-term, high doses of nicotine is rare. Of much greater concern is long-term exposure through habitual smoking or chewing tobacco. Annually, over 435,000 deaths in the United States and 2.5 million deaths worldwide are attributed to tobacco smoking. The main illnesses caused by long-term tobacco use are cancer, lung disease, and heart disease. Cigarette smoke contains cancer-causing chemicals. One such chemical damages a human gene that suppresses cancer growth in the body.

Lung problems caused by cigarette smoking include difficulty breathing, wheezing, chest pain, congestion, and lowered resistance to infections. Tobacco is also the leading cause of emphysema, which is a failure to clear the lungs, making breathing difficult or impossible. Smoking increases blood pressure and heart rate. It also increases hardening of the arteries (atherosclerosis) and obstruction of blood vessels (thrombosis). Smoking increases the odds of death by heart disease by three to nineteen times.

Prenatal and neonatal effects. Smoking during pregnancy is associated with reduced body size during the fetal stage and infancy. The child also has a greater tendency toward obesity by age five. Passive exposure to smoke impairs breathing in children, especially in those with respiratory illnesses such as asthma. Cigarette smoking during pregnancy increases the risk of sudden infant death syndrome.

Areca

Areca *(Areca catechu)* is a palm tree that grows in India, southeast Asia, the East Indies, and East Africa. The seed of the areca tree is chewed for its stimulant effects by more than 200 million people, particularly in Asia. It is the fourth most popular drug in the world, following nicotine, alcohol, and caffeine.

The areca nut is sometimes erroneously called the betel nut. A *betel quid* is the areca nut and lime (calcium hydroxide) rolled into the leaf of the betel pepper *(Piper betle)*, which is chewed and kept in the mouth.

Active Chemicals in Areca

A few active drugs are found in areca: **arecoline,** arecaidine, guvacoline, and guvacine. However, arecoline is the most abundant and most potent, and is probably responsible for creating the majority of the psychological effects.

How Areca Works

Arecoline and the other drugs in areca stimulate the acetylcholine system in the brain. Whereas nicotine does so through nicotinic receptors, arecoline does so through a different receptor, the muscarinic receptor. So through a slightly different means, the brain is tricked into acting as if it had more acetylcholine, leading to the stimulant effects. Also like nicotine, arecoline increases dopamine activity in the brain, which is related to its addicting effects.

Effects of Areca

Autonomic effects. Areca causes sweating, facial flushing, and a warm sensation in the skin. This is caused by regulation of blood vessels through the autonomic nervous system, bringing more blood to the skin. Heart rate and blood pressure are raised by chewing areca.

Brain activity. As expected from a stimulant, areca increases energy use and blood flow in the brain. This is especially true in parts of the brain that use the neurotransmitter acetylcholine. Consistently, areca chewing also activates electrical activity in the brain.

Cognitive effects. Animal and human studies both show that arecoline improves memory. Humans given arecoline process information faster. Arecoline has been tested in people with Alzheimer's disease, with improvement noted in several areas of mental function, including memory. The results have been mixed, however, in different studies. Arecoline is not likely to become a standard treatment for Alzheimer's disease. Drugs such as Aricept and Exelon accomplish the same results, probably more effectively.

Addiction and Dependence

As with tobacco, many who chew areca develop an addiction to it; it stimulates the acetylcholine system and indirectly affects dopamine, an important neurotransmitter in addiction. People often continue to use areca even though they experience noticeable negative effects on their health, which is one of the characteristics of addiction.

Side Effects

The most displeasing effect of chewing areca nuts is a reddish-brownish discoloration of the mouth and gums and a tendency toward oral skin peeling. This has been called betel chewer's mucosa, and other oral lesions such as leukoedema, leukoplakia, and ulceration may occur. The teeth also become stained reddish-brown or black in the most extreme cases. Chewing areca also increases the likelihood of oral cancers.

Areca is associated with cardiovascular disease and worsening diabetes and asthma. Areca also interacts with antipsychotic medications for schizophrenia. Cases have been reported of schizophrenic patients who developed severe motor symptoms after chewing areca while on medication.

Lobelia

Lobelia *(Lobelia inflata)* is an annual or biennial flowering herb. It was named after a Flemish botanist, Matthias L'Obel. It is also referred to as Indian tobacco since Native Americans used it as a substitute when tobacco was unavailable.

Active Chemicals in Lobelia

Lobelia contains many drugs, but the most active is **lobeline.** Similar to nicotine, lobeline activates nicotinic receptors for acetylcholine in the nervous system. However, it works as a "partial agonist" in this regard, activating the receptor but not to the full extent as tobacco does. For this reason, lobelia lozenges have been tried as a treatment to help people quit smoking, with poor results.

Effects of Lobelia

Lobelia sometimes produces feelings of choking, pressure in the throat and chest, and a dry cough. It seems to have reinforcing effects, similar to those of tobacco. However, it has not been well studied in humans.

AMPHETAMINE AND AMPHETAMINE-LIKE STIMULANTS

Three stimulant plants have amphetamine-like effects. This is not a coincidence, but occurs because they work through the same neurotransmitters: dopamine, norepinephrine, and serotonin. In fact, although amphetamine is a synthetic drug, it was designed with **ephedrine** in mind, the first of the three stimulants in this section.

Ephedra

Ephedra is a group of related plant species that have long been used for their stimulant effects. One member of this group is *Ephedra sinica,* a shrub with long, straight branches and small, pointed leaves. *E. sinica* is native to Asia, growing mostly in Mongolia and neighboring parts of China.

Ephedra is also known by the Chinese name, ma huang, and has been used by the Chinese for more than 5,000 years. A similar species, *Ephedra nevadensis,* grows in North America. The Mormons were introduced to it by Indians when they arrived in Utah. Since their religious beliefs prohibited them from drinking tea and coffee, they made tea from ephedra instead, and it became known as Mormon tea. Ironically, it's a more powerful stimulant than coffee or tea.

Ephedra might be the oldest known human stimulant, since remains of the plant were discovered in a 50,000-year-old Neanderthal grave in Iraq. Modern medicine uses chemicals from ephedra, which began with identification of the chemical ephedrine in 1887.

Active Chemicals in Ephedra

Ephedra contains several active chemicals, with ephedrine being the most significant. *Ephedra sinica* contains 1.25 percent ephedrine.

Another chemical, pseudoephedrine, is made synthetically and marketed as an over-the-counter decongestant (e.g., Sudafed).

Drug Issues for Ephedra

When ephedra is bought over the counter, it is commonly dried, ground, and packaged in a capsule. A recent analysis of such capsules (which vary from brand to brand) showed that 400 mg of ephedra contained about 5 mg of ephedrine and 1 mg of pseudoephedrine. The half-life of the drug is about five hours, and when compared to pharmaceutical tablets of ephedrine (the kind that would be prescribed by a physician) it was very similar. The "herba" form absorbs slower, so the effects take longer to begin, but both had the same effects.

How Ephedra Works

Ephedrine imitates the neurotransmitters epinephrine and norepinephrine (also known as adrenalin and noradrenalin). It directly activates the receptors for these neurotransmitters, which are called alpha and beta receptors. Many people are familiar with the blood pressure medications called beta blockers (such as Inderal). Ephedrine does the opposite of these medications. Beta blockers (as the name implies) block the receptor, preventing its activation, while ephedrine attaches and activates the receptor.

Effects of Ephedra

Cardiovascular. As you might expect, ephedrine increases heart rate and increases systolic blood pressure.

Lungs. Stimulation of beta receptors in the lungs by ephedrine relaxes bronchi. This allows more oxygen to come into the blood and carbon dioxide to pass out. This normally happens when we are physically active, but ephedra creates this effect without physical activity. For this reason, ephedra has long been used to treat asthma.

Psychological effects. Ephedra potently stimulates the nervous system. As a result, the person feels alert and active. It also lifts a person's mood temporarily, making him or her feel more optimistic and full of energy. Depending on the dose and when it's taken, the effects may be excessive, causing temporary anxiety and insomnia. On tests of attention, ephedrine partially reduces the impairment caused by sedating drugs such as antihistamines. Pseudoephedrine has similar stimulant effects, but is less potent.

Side Effects

The most common side effects of ephedra include insomnia, anxiety, and cardiovascular stimulation. Normal doses range from 15 to 30 mg of actual ephedrine, and the lethal dose is in the 1 to 2 g range. Even small overdoses can have very negative effects such as headache, irritability, restlessness, anxiety, insomnia, rapid heartbeat, urinary disorders, and vomiting. When large doses are consumed for a long time "toxic psychosis" may occur, which is a temporary inability to discern reality. Users may experience hallucinations and paranoid delusions.

Ephedra should not be used by anyone with high blood pressure or any form of heart disease, unless directed by a physician. Cases of people having strokes after taking ephedrine have been reported. The risk of stroke increases when people combine ephedrine with other stimulants such as the over-the-counter stimulant phenylpropanolamine. In contrast, pseudoephedrine seems to be safer in this respect. A large sample (more than 100,000 people) of people taking over-the-counter pseudoephedrine found no cases of stroke within fifteen days of taking it.

In the past few years, ephedra has been sold with misleading names such as "herbal ecstasy," alluding to the club drug Ecstasy (MDMA). Several deaths have resulted, mostly from high blood pressure and abnormal heart rhythms.

As with most other stimulants, ephedra probably aggravates anxiety disorders. It is not really an effective treatment for depression since its mood-elevating effects are temporary. Ephedra should also not be taken by people who have glaucoma, prostate problems, or excessive thyroid hormones. Medications that may interact adversely with ephedrine include heart glycosides, halothane, guanethidine, MAO inhibitors (antidepressants), secale alkaloids, and oxytocin. Pursuant to these potential health risks, in December 2003 the FDA announced a ban on over-the-counter sales of ephedra in the United States. Although sales of the herb made up only a small portion of overall dietary supplement sales, it accounted for the majority of reported adverse side effects. Furthermore, ephedra has been associated with as many as 164 deaths in the United States. Given the seriousness of its side effects and its potential for improper use, the nonprescription use of ephedrine in the form of ephedra supplements was deemed unsafe.

Addiction and Dependence

Use of ephedrine is sometimes associated with dependence. Most people who take it do not abuse the drug, but as with any stimulant, it has some abuse potential.

Coca

Coca *(Erythroxylon coca)* is a small shrublike tree, with tough, oval green leaves. The coca plant is native to the Andes region of South America, but it is also cultivated in Indonesia, India, and Sri Lanka. Many related *Erythroxylon* species grow throughout tropical regions of South America.

Coca is probably the most infamous of the stimulant plants. Use in the United States peaked in the 1980s; 17.3 percent of high school seniors in 1985 reported using it at least once in their lifetimes. This number fell to a low of 5.9 percent in 1994, but rose again to 9.8 percent in 1999.

Coca has long been used by Indians in the Andes, who chew the leaves for stimulation and endurance. Coca was an important part of the culture of the Incas, who regarded "Mama Coca" as a goddess. Knowledge of the plant spread to Europe with colonialism, but widespread use really began after the drug **cocaine** was isolated in 1855. Coca-Cola contained a mixture of extracts from the coca plant and kola nut, and a French wine named Vin Mariani also contained coca oil. At the turn of the twentieth century in the United States, the euphoric effects of cocaine led people to mistakenly assume it was a health tonic that increased strength and well-being. After a time, however, the negative effects became evident and laws were enacted to control its use.

The white powder form of cocaine (cocaine hydrochloride, or cocaine HCl) is prepared by mashing the leaves with a solvent. When the solvent evaporates, a coca paste is left, containing the essential chemicals from the leaves. In South America, this paste is sometimes directly mixed with tobacco and smoked. The paste can be further refined to extract pure cocaine. Cocaine HCl is the cocaine molecule attached to a salt, which makes it stable and soluble in water. As such, it can be snorted and absorbed through the lining of the nose, the nasal mucosa. Cocaine HCl can also be made into a "freebase" form by mixing it with the solvent ether. In freebase form it can be smoked, but the flammability of ether makes this a dangerous practice. However, it was discovered that freebase cocaine could be made more easily and safely by mixing it with baking soda and water, a compound commonly known as crack.

Active Chemicals in Coca

As with most plants, coca contains a number of active chemicals. However, cocaine is the main drug and by itself can account for all of the psychological effects associated with the plant.

Drug Issues with Coca

Cocaine is taken in several ways, and the way it is taken has a major impact on its effects. The leaves may be chewed, which makes absorption slow and gradual. This produces a mild, prolonged high, probably approximating the intensity of strong espresso (see Figure 4.4). Snorting cocaine makes it absorb faster through blood vessels inside the nose. Cocaine constricts blood vessels, so it doesn't absorb very well this way.

Cocaine is less frequently injected intravenously, but is sometimes combined with heroin to make an especially addicting concoction called a speedball. Cocaine and heroin both cause euphoria, so taking them together magnifies the euphoric effect. On the other hand, cocaine causes stimulation and heroin causes sedation, so they can counteract each other's negative side effects. As a result, this is one of the most potent and addicting combinations of drugs. Unfortunately, effects on the heart are not altered, and users sometimes experience heart attacks or strokes.

Smoking crack is the fastest and most direct means of delivering cocaine to the brain, and thus probably the most addicting. When equal doses of cocaine are smoked or injected intravenously, the

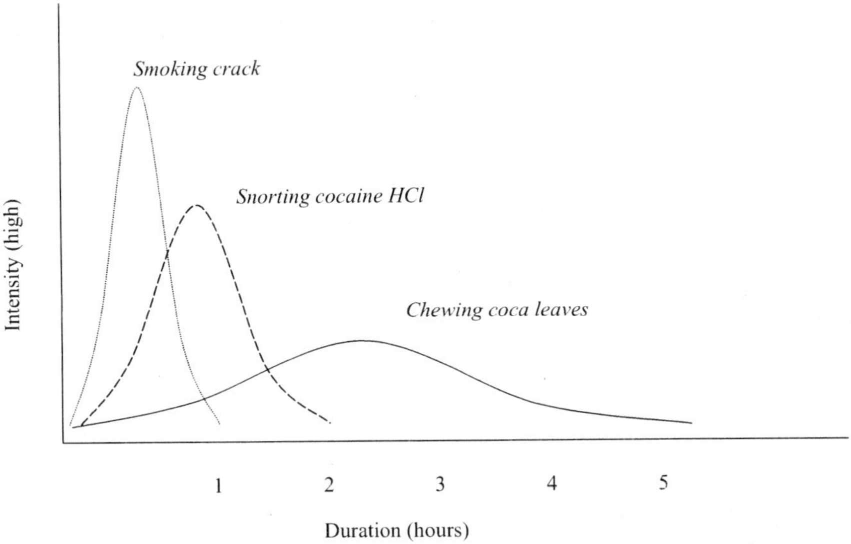

FIGURE 4.4. Comparison of smoking crack, snorting cocaine, and chewing coca leaves. Crack causes the fastest, most intense effects.

smoked cocaine produces the most intense responses. Crack users describe the experience as unbelievably euphoric, using terms such as "full-body orgasm." When smoked, cocaine is very rapidly and completely absorbed through the inner surface of the lungs. Since blood from the lungs goes directly to the brain, the effects can be felt within seconds.

Cocaine is rapidly broken down in the blood and liver, with a half-life of thirty to ninety minutes. Although cocaine itself is detected in urine for only twelve hours, the breakdown products can be detected for forty-eight hours to two weeks. When consumed with alcohol, a breakdown product of cocaine (called benzoylecgonin) combines with the alcohol to make another chemical called cocaethylene. Cocaethylene has effects similar to active cocaine, so taking cocaine and alcohol together magnifies the effects of the cocaine.

How Cocaine Works

Cocaine's stimulant effects are a result of its effects on the neurotransmitters dopamine, serotonin, and norepinephrine. Cocaine prevents neurons from taking up these three neurotransmitters after they are released, prolonging their effect. As a result, they remain in the synapse longer and activate other neurons (see Figure 4.5). Chronic use of cocaine causes adaptation in the receptors for these neurotransmitters, as the brain attempts to compensate for the extra activity. The receptors become less sensitive, and if the drug is discontinued, the brain is then underactivated. This underactivation is the cause of cocaine cravings and withdrawal symptoms, leading to relapse and addiction.

Effects of Cocaine

Psychological. Cocaine causes intense mental stimulation. The user becomes very active and talkative. There may be a "flight of ideas," which is a rapid, disorganized stream of thoughts. Emotion-

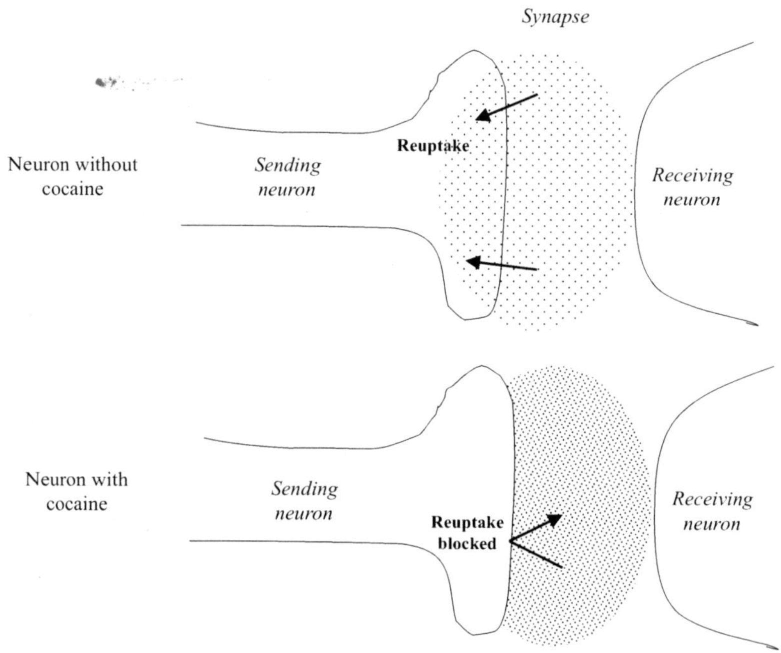

FIGURE 4.5. Cocaine in the synapse. Normally, without cocaine, the neurotransmitters are taken back into the neuron and inactivated. With cocaine, reuptake of neurotransmitters is blocked, keeping them active longer.

ally, cocaine causes euphoria, lifting mood to above-normal heights. The user experiences inflated self-esteem and may overestimate his or her abilities. Tiredness and need for sleep are reduced. Cocaine is also a powerful appetite suppressant.

Although short-term use of cocaine wakes up the brain, long-term use has been associated with problems in mental abilities such as concentration, memory, and executive functions (e.g., planning, reasoning, and judgment). However, we cannot be certain that cocaine *causes* these problems, only that there is some relationship. Cocaine may alter and damage areas of the brain that control those functions. On the other hand, it may be true that people who have these problems to begin with are more likely to use cocaine. Either one is possible, and the reality may be a combination of both.

Autonomic effects. Cocaine stimulates the sympathetic branch of the autonomic nervous system. This has the effect of increasing heart rate and blood pressure, and dilating bronchi in the lungs.

Brain activity. Brain imaging studies show that cocaine increases energy use by the brain, as would be expected from a stimulant. However, while most other stimulants increase blood flow to the brain, cocaine decreases it because it causes blood vessels to constrict. These changes in brain activity are directly related to euphoria and cravings.

Addiction and Dependence

Addiction is a significant problem with cocaine use. Cocaine creates intensely rewarding effects. These effects seem to be especially mediated by the neurotransmitter dopamine. In animals and humans alike, a binge pattern of use typically occurs, so that it is repeatedly used over a period of hours or days until the supply is used up.

The cocaine withdrawal syndrome mirrors the effects of the drug. It includes depression, sleepiness, fatigue, increased appetite, slow heart rate, and craving for cocaine. Treatment of cocaine addiction is a stubborn problem because the effects of the drug are so intense and pleasurable. Effective treatment requires some form of social support and approval for abstaining (e.g., Narcotics Anonymous). Drug treatments of cocaine addiction have been attempted with antidepressants and other related drugs, with limited success. A new approach is being researched, involving antibodies for cocaine. This strategy induces the body to react to cocaine and neutralize it in the bloodstream, making it less effective or ineffective.

Cocaine Toxicity

Cocaine alters heart rate and can result in abnormal heart rhythm and heart attack (cardiac arrest). Cocaine constricts blood vessels and raises blood pressure, which can cause a stroke, or lack of blood supply to the brain leading to permanent damage. It's not uncommon for people in their thirties to experience heart attacks and/or strokes resulting from cocaine use.

Cocaine crosses the placenta and affects fetuses. The effect of cocaine in prenatally exposed children is another tricky research question, since many other factors are in the picture (e.g., use of other drugs, nutrition, parents' mental abilities, etc.). However, cocaine given to rats does cause behavioral problems. In humans, this probably results in attention deficits and impulsivity.

Heavy use of cocaine can causes a toxic psychosis, giving users hallucinations or paranoid delusions. They may feel someone is "out to get them" and may become violent. Some report a tingling feeling that is interpreted as bugs crawling under the skin ("coke bugs"). This condition resolves once use is discontinued, but the person may become more sensitive to this effect of cocaine with future use. A small percentage experience lasting psychotic problems after using cocaine, but these are probably people who had existing problems that were exacerbated by cocaine.

Khat

Khat *(Catha edulis)* is a shrub or tree native to East Africa, and is cultivated in countries such as Ethiopia and Yemen. It has long been known in the Middle East, and Arabic writings mention it as far back as 1237 A.D. In some Muslim cultures, khat is accepted socially and regarded as an acceptable alternative to alcohol. In Yemen, users gather in someone's home to chew the leaves and converse. Some estimate that 5 million portions are used daily. Use in the West is less common, in part because the leaves are perishable. However, methcathinone (or "cat"), an amphetamine-like synthetic drug that was derived from khat, is known to be abused in the West.

Active Chemicals in Khat

Several drugs in khat have stimulant effects. The main drug, however, is a chemical called **cathinone,** which accounts for the majority of effects.

How Khat Works

Khat is usually consumed by chewing the fresh leaves and swallowing the juices. Since it is consumed by chewing, the absorption of cathinone is slow and prolonged. The psychological effects are maximal at about two hours, and last for four hours. As with chewing coca leaves, this gives khat a mild, prolonged effect compared to a drug that is injected or smoked. When cathinone is taken in these ways, the effects are more sudden and intense.

Khat works through the same neurotransmitters as cocaine: dopamine, norepinephrine, and serotonin. It seems to work almost identically to amphetamine in that it causes leakage of these neurotransmitters out of the neuron. Their increased presence activates other neurons and has an overall stimulant effect.

General Effects

Khat's effects are very similar to those of other amphetamine-like stimulants, that is, increase in mental stimulation and physical endurance, lifted mood, and appetite reduction. When users take khat in the traditional social setting, talkativeness and sociability increase. Khat also similarly stimulates the sympathetic nervous system, increasing heart rate and blood pressure.

Addiction and Dependence

Khat has abuse potential, as do other similar stimulants. The pattern is similar in both animals and humans. Withdrawal symptoms of khat include tiredness, depression, nightmares, and mild tremors. Chewing khat, versus injecting or smoking it, probably limits its addictiveness. Since the drug is absorbed slowly by chewing, and it is fairly quickly eliminated, large amounts don't accumulate in the body, so the khat chewer never gets a huge dose. This is another key example of the effects of a drug being altered by the way it is taken.

Side Effects

As with other amphetamine-like stimulants, the main side effects come from overstimulation of the sympathetic nervous system, resulting in raised blood pressure and accelerated heart rate. These effects could result in heart attack or stroke in a vulnerable individual. Chronic overuse of khat has been reported to cause headaches, anorexia, insomnia, constipation, liver disease, and respiratory illnesses. Some chronic users report experiencing hallucinations just before they fall asleep. Very heavy chronic use can cause a toxic psychosis.

SUMMARY

As we have seen in this chapter, several plants are known for their mental stimulant effects. They all work through a variety of chemical systems in the brain. It is interesting to note that stimulants that work through similar brain chemicals also have similar psychological effects. For example, cocaine and ephedra affect similar neurotransmitters and thus have similar psychological effects. However, tobacco affects a different neurotransmitter, and has somewhat different stimulant effects.

All of the stimulants have some potential for abuse. This varies with social and environmental influences, in what context people use the drug, and also their inherited predispositions. Still, the chemical effects of the drug are highly important. Fairly consistently, the stimulants that have a strong effect on the neurotransmitter dopamine tend to have more potential for abuse and addiction.

Whether we use them for positive or negative purposes, it is likely that our relationships with stimulant plants will long continue.

Chapter 5

Cognition-Enhancing Herbs

We humans rely heavily on our **cognitive,** or thinking, abilities to survive. It's the thing that distinguishes us the most from all other species on the planet. Rather than operating on instincts and reacting to our immediate circumstances, we tend to rely more on long-term goals and mental concepts. This shift in strategy to mental skills was accompanied by changes in the human brain: our cortex is proportionately enlarged. A particular part of the cortex that is larger in humans is the frontal lobes. It's a critical brain structure in many mental functions, and when it's damaged or undeveloped, people tend to function very inefficiently.

As a consequence, in our search for plant medications, we pay special attention to any that improve mental abilities, even if only temporarily. Any plant that might enhance those abilities is a valuable commodity.

DISCUSSION AND TREATMENT OF DEMENTIA

One use of mind-enhancing drugs is to treat people with **dementia,** which is a general term for a loss of mental abilities. It may be caused by several different illnesses, and may present in many different forms. The financial cost of health care for dementia is enormous, since people so afflicted require progressively greater amounts of supervision and eventually need total care. The cost of human suffering, of both the afflicted and their families, is unquantifiable.

Dementia generally involves an impairment in memory and other mental abilities, which causes difficulty functioning in daily life (e.g., social functioning, self-care, ability to work). The *Diagnostic and Statistical Manual of Mental Disorders,* published by the American Psychiatric Association (current edition, DSM-IV-TR), is the standard manual for identifying mental illness. It defines dementia as a loss of memory and at least one other mental ability, such as language, learned abilities, recognizing objects or people, and executive functions (organization, planning, abstract thinking, etc.). However, the symptoms depend heavily on the type of dementia a person has. Some forms cause only minor memory problems.

Dementia has both degenerative and nondegenerative forms. Degenerative forms involve a progressive loss of neurons, and the brain must work progressively harder to compensate. Eventually it can't compensate, and a person's mental abilities decline. Nondegenerative forms may not progress any further. Another thing to consider is the speed of onset. Most dementias have a slow and gradual onset, worsening over a period of weeks or months. If there is a sudden onset (e.g., overnight), either it's a problem with blood flow to the brain or it's some illness other than a dementia (such as a stroke). Another problem called **depressive pseudodementia** may seem like a dementia but is really a form of depression in an elderly person. The main difference is that once the depression is treated, the condition improves. Unfortunately, the same cannot be said for most other forms of dementia.

Normal Aging

Before we discuss dementia, it is important to understand the mental changes in normal aging. In a healthy older person, there is some reduction in brain weight—about 15 percent across the life span, and about 2 percent per decade after the age of sixty. Metabolism and activity are also reduced in certain regions of the brain with aging. Along with this come age-related changes in mental abilities. Elderly people have relatively more difficulty in the speed at which they can process information. They also have some impairment in short-term memory. Although it may be harder for them to form new memories, their memory for familiar information and skills is well preserved. Literacy and education play an important role in preserving mental ability. They cause the creation of new connections among brain cells, which act as a reserve to be drawn upon as the brain ages. Learning can continue throughout the life span, and new connections can constantly be made, keeping the mind as young and flexible as possible.

It cannot be stressed enough that in the absence of any disease (e.g., Alzheimer's disease, hardening of the arteries in the brain) a person's mental abilities will remain mostly intact throughout life. Many people maintain a mentally active life well into their nineties. The term "senile dementia" is highly misleading, since getting old in and of itself does not cause significant mental decline. Disease impairs the mind, not simple aging.

Alzheimer's Disease

The most common form of dementia is Alzheimer's disease, which accounts for about 20 percent of all psychiatric hospitalizations. The odds of developing it increase as a person gets older: the risk between the ages of thirty and fifty-nine is 0.02 percent, but the odds increase by ages eighty to eighty-nine to 10.8 percent.

The disease progresses through three general stages, which last two years each on average. The first stage is characterized by difficulties with memory for recent events or learning new information (such as phone numbers and names). Language problems occur, such as remembering the name of something familiar. These are common problems for normally aging people, and they sometimes cause unnecessary fears of developing Alzheimer's disease.

The second stage is more severe, and more mental abilities are affected. In this stage, people need more and more help in their daily lives, since they may have difficulty doing once-familiar things. Instead of having memory problems only for recent events, they start to lose earlier memories as well. People also may change during this stage, seeming relatively well at times and severely affected at others. In the final stage, victims lose all higher mental abilities. They cannot communicate, form thoughts, or recognize familiar people. They enter a vegetative state in which they are bedridden and unable to perform basic self-care. Death results from health complications, such as infections from bedsores or a weakened heart, and not directly from the disease itself.

Risk Factors for Alzheimer's Disease

Although the odds of getting Alzheimer's disease increase with age, it is important to remember that it is a disease and not a part of normal aging. It is to some degree inherited, but at present that only explains a fraction of the cases. A gene for a certain form of the protein ApoE **(apolipoprotein E)** has been associated with an increased risk for Alzheimer's disease. Still, even people with this abnormal form of the gene don't have a 100 percent chance of developing the disease.

Another risk factor for Alzheimer's disease appears to be lower levels of education. As mentioned, learning increases the connections in the brain. These can help the brain compensate for the loss of neurons that occurs.

Biological Basis of Alzheimer's Disease

When the brain of a person with Alzheimer's disease is looked at under a microscope, certain telltale signs of the disease can be detected. These are mainly called **neuritic plaques** and **neurofibrillary tangles** (or simply plaques and tangles). The plaques are abnormal accumulations of **beta-amyloid** and ApoE proteins. The tangles are tangled-up extensions of neurons that no longer function. Plaques and tangles actually are found in all people as a consequence of aging. However, a person with Alzheimer's disease will have a great excess.

Drug Treatments for Alzheimer's Disease

Alzheimer's disease is also associated with a loss of the neurotransmitter acetylcholine. Acetylcholine is important in the formation of memories, so it is not surprising that memory dysfunction is one of the earliest symptoms. Current treatments for Alzheimer's disease are geared toward boosting acetylcholine in the brain, to make up for the lack. This works to a limited degree, but it does not stop the progress of the disease.

The drugs that do this and are currently available are donepezil (Aricept), rivastigmine (Exelon), and galantamine (Reminyl). The first of these drugs, tacrine (Cognex), has been associated with heart and liver problems, so it's no longer used. As would be hoped, these drugs increase acetylcholine levels in the brain by preventing breakdown. As a result, the brain is more active and a small but measurable improvement occurs in mental functions. These drugs do not stop or reverse the disease; they only attempt to treat the main symptom.

Another strategy for Alzheimer's disease has more of a preventive nature. Antioxidants are nutrients that neutralize toxic chemicals in the brain known as free radicals. This protects the living neurons from damage done by the free radicals that would otherwise kill them. It has been shown that elderly people with high blood levels of antioxidant vitamins (namely vitamins A, C, and E) have better mental functions than people with low levels.

Two medications have been tested in people with Alzheimer's disease for their protective effects against free radicals: vitamin E and selegiline (Eldepryl). Vitamin E is a very effective antioxidant, and people with Alzheimer's disease have below-normal levels of it in their nervous systems. One study has shown that either vitamin E or selegiline can slow the progression of Alzheimer's disease and delay the necessity for nursing home placement.

Ironically, **nonsteroidal anti-inflammatory drugs (NSAIDs),** such as **aspirin,** have been evaluated for their protective effects against Alzheimer's disease. It was discovered that people regularly taking NSAIDs have a 50 percent lower chance of developing Alzheimer's disease. It's not certain how this works, but the disease seems to have an inflammatory stage when neurons die, so anti-inflammatory drugs may prevent this.

Vascular Dementia

Vascular dementia is another very common form of dementia. As the name suggests, it happens whenever blood flow to the brain is compromised, which can happen for multiple reasons. The factors that place a person at higher risk for vascular dementia are high blood pressure, hardening of the arteries (atherosclerosis), high alcohol consumption, stress, and exposure to toxic chemicals. As with Alzheimer's disease, higher levels of education may protect a person to some degree. In one form of this illness, the brain suffers multiple microscopic strokes deep in the white matter, making communication from one part of the brain to another more difficult.

GINKGO

History and Botany

Ginkgo biloba is also known by the names kew tree or maidenhair tree. It is an ancient tree that has been in existence for 200 million years, and some living trees are estimated to be over 3,000 years old. The **ginkgo** tree grows to a height of 35 to 80 feet and has characteristic green, fan-shaped leaves (see Figure 5.1). Its seeds are round and fleshy, having a yellow or green color. It is a popular herbal supplement, and over 50 million ginkgo trees are cultivated for the preparation of supplements.

Ginkgo is a Japanese word that comes from the Chinese name *yinshing,* meaning "silver apricot." Ancient Chinese medical writings recognized ginkgo as a treatment for memory loss, among other ailments. The ginkgo tree was brought to Europe and North America in the 1700s, but medicinal use of ginkgo did not become popular in the West until the 1960s. Now it is among the most commonly prescribed drugs in France and Germany, making up 1 percent and 4 percent of their prescriptions, respectively.

Active Chemicals in Ginkgo

Ginkgo's active chemicals are classified in two main groups: **flavonoid glycosides** and **terpene lactones.** Standardized extracts of ginkgo ensure that a certain amount of both of these classes of chemicals are present. One common extract contains 24 percent flavonoid glycosides and 6 percent terpene lactones. Use of ginkgo as a tea may be ineffective, since many of these chemicals do not diffuse well into water.

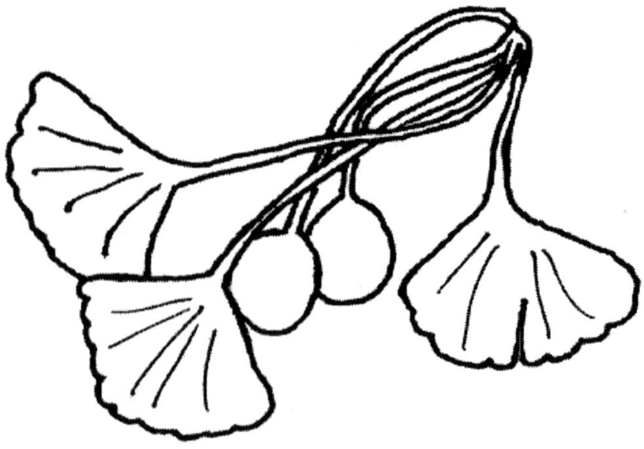

FIGURE 5.1. Ginkgo *(Ginkgo biloba).* Copyright 2004 by Marcello Spinella.

How Ginkgo Works

Ginkgo may work in several ways in the brain. Exactly how it works is not yet known, but a combination of effects may be responsible. Ginkgo boosts activity of the neurotransmitter acetylcholine in the brain. It increases the amount of acetylcholine released by neurons, and it increases acetylcholine receptors, making the neurons more sensitive to acetylcholine's effects. Acetylcholine is absolutely necessary for the formation of new memories. People with Alzheimer's disease lack acetylcholine, and boosting it improves their memory.

Ginkgo also has antioxidant effects and may protect neurons from free radical oxidative damage. Oxidative damage is very harmful to neurons, damaging the components that are needed to survive (such as their DNA), making them eventually decline and die. Oxidative damage is excessive in many degenerative brain diseases such as Alzheimer's disease, and as stated previously, antioxidants such as vitamin E seem to slow down the process. It seems likely that ginkgo has a real antioxidant benefit for people with Alzheimer's disease, but the research has yet to be done. Oxidative damage at a slower rate may be responsible for the effects of normal aging, so it's also possible that ginkgo may help a healthy brain remain that way longer.

Drug Issues with Ginkgo

Ginkgo absorbs well into the body when taken orally. Blood levels of the active chemicals in ginkgo peak around three hours after taking it. Given the amount of time it takes the body to process ginkgo, taking it two to three times per day would allow a person to keep steady blood levels. Ginkgo does not seem to adversely affect the liver.

Effects of Ginkgo

Vascular System

Ginkgo relaxes the small muscles around blood vessels, increasing blood flow, particularly in small blood vessels, and lowering blood pressure. Ginkgo also inhibits a chemical in the body called platelet-activating factor (PAF), so it reduces blood clotting. Since it reduces blood clotting, ginkgo may prolong bleeding time. Long-term treatment with ginkgo (three months) at certain doses does reduce diastolic blood pressure. Ironically, a smaller dose (120 mg/day) had this effect, but a higher dose (240 mg/day) did not. Since ginkgo can affect bleeding and blood pressure, people should inform their doctors of their desire to take ginkgo, especially if they are taking blood thinners, blood pressure medication, or having surgery.

Sexual Effects

There has been speculation that ginkgo can enhance sexual function. Since it seems to increase blood circulation, this is possible since sexual arousal is accompanied by increased blood flow to the genitals. Medications such as Viagra work entirely by increasing blood flow. One study, which was not controlled, suggested that ginkgo relieved some of the sexual dysfunction caused by antidepressant drugs. Although this is an encouraging report, controlled studies are needed to see if this effect is real and worthwhile.

Inner Ear Effects

Ginkgo has been tested for effects on dizziness and tinnitus (ringing in the ear). The results of studies of its effectiveness for tinnitus have not been consistent, and a recent study suggests that it doesn't work. Some research suggests that it may help to alleviate dizziness and vertigo, but more work needs to be done.

Brain Activity

Ginkgo seems to increase blood flow to the brain in rats and humans. However, it is not known if this effect persists when ginkgo is taken over a long period of time. Ginkgo also seems to increase electrical activity and energy use in the brain, which would be expected from a drug that activates the brain. Ginkgo causes electrical activation in the brain similar to the Alzheimer's disease medication Tacrine.

Hormonal Effects

Ginkgo seems to reduce levels of stress hormones and the learning impairments that result from stress. It does this by reducing stress hormone production in the adrenal glands. High levels of these hormones damage the brain, particularly areas responsible for memory. Consequently, conditions that cause high levels of stress hormones (such as chronic stress, chronic depression, post-traumatic stress disorder [PTSD], and Cushing's disease) also cause impairments in memory. So it is possible that ginkgo could protect memory from the damaging effects of stress hormones.

Cognitive Effects

Both animal and human studies of ginkgo suggest that it improves certain mental abilities. Animals given ginkgo learn faster and have better short-term memory.

Many studies have tested ginkgo in humans with dementia. The best research has been done on Alzheimer's disease. An analysis of many well-designed studies showed that ginkgo is better than a placebo in the treatment of Alzheimer's disease. The best dosage is difficult to determine since different studies use different doses. However, the best effects were found at 120 to 240 mg per day. The benefit was by no means miraculous, but no miraculous treatment exists for Alzheimer's disease. In fact, ginkgo seemed to be as effective as the prescription medication Aricept. Similar to other medications, the actual benefits are realized after the medication has been taken for a few weeks.

Some research has been conducted on the effectiveness of ginkgo in other illnesses such as age-related memory decline and vascular dementia, but more work needs to be done in these areas.

Of more interest to the average person are the effects of ginkgo in healthy people. A few studies show that ginkgo may actually help healthy people to some degree. Improvements have been seen in memory and executive functions. Again, it may be necessary to take ginkgo for a few weeks before the full effects kick in.

Side Effects

No serious side effects were reported in the large clinical studies of ginkgo. The main side effects are gastrointestinal disturbance, headaches, dizziness, and palpitations. However, some studies have reported that the side effects were the same whether actual ginkgo or a placebo was given.

The most troublesome potential side effect of ginkgo is its effect on blood clotting. A few cases of bleeding in or around the brain have been associated with ginkgo. However, most of these instances involved a person taking ginkgo along with another blood thinner (such as aspirin or warfarin). Individuals with blood clotting disorders (such as hemophilia), taking a blood thinner medication, or having surgery should consult a doctor regarding the use of ginkgo.

GINSENG

Ginseng has a long history of use in several cultures. The name *ginseng* has been used to refer to several different plants, so it's important to keep the names straight. Asian ginseng refers to *Panax ginseng,* which grows in China, Korea, Japan, and Russia. *Panax* comes from the Latin words meaning "cure-all," since ginseng is thought by some to treat many different illnesses. Ginseng may be either white or red, depending on how it's prepared: white ginseng is simply the peeled and dried root, while red ginseng is steamed and then dried. This is potentially important, since different chemicals are produced by the steaming process.

American ginseng is *Panax quinquefolius* (see Figure 5.2), which is very similar to Asian ginseng. It grows in several parts of the

FIGURE 5.2. American ginseng *(Panax quinquefolius).* Copyright 2004 by Marcello Spinella.

United States, and most of it is from Wisconsin. Siberian ginseng *(Eleutherococcus senticosus)* is not truly ginseng, but is so named only because it was thought to have some similar effects. Little research has been done on Siberian ginseng, so it is not discussed here.

Ginseng is a small perennial plant, growing 30 to 80 cm high. It produces green leaves, a red pea-sized fruit, and flowers. The underground stem (or rhizome) is the part used medicinally. Ginseng is very expensive to produce, requiring about a $20,000 investment to plant and harvest one acre. Relatively few seeds produce actual plants, and it should be harvested after six years. As a result, pure American ginseng can sell for $400 to $600 per pound. For farmers, the money is significant, but so are the risks. The United States is a major exporter of ginseng, and has exported it to Asia since the eighteenth century. Since it has been extensively harvested in the wild, ginseng is an endangered species in the United States and must be harvested responsibly.

Active Chemicals in Ginseng

Ginseng contains a mix of active chemical constituents that create its effects. These are known collectively as **ginsenosides.** They are all very similar in chemical structure and are classified as steroids.

How Ginseng Works

Ginseng has numerous effects in the brain. Exactly how it works is still unknown, but there are a few good theories. Ginseng increases the release of acetylcholine in the brain. It may also directly stimulate acetylcholine receptors. Since acetylcholine is so important for memory, ginseng could improve memory. Ginseng also dilates blood vessels, increasing blood flow. So it could also increase blood flow to the brain. Ginseng may also have antioxidant effects, protecting neurons from free radical damage.

Drug Issues with Ginseng

Ginseng is rapidly absorbed when taken orally. Levels of ginseng chemicals peak in the blood at about thirty minutes. Although most drugs are broken down by the body, ginseng chemicals are not, so they are excreted in the same condition they were taken.

Effects of Ginseng

Cardiovascular Effects

Ginseng causes a slight decrease in heart rate. At normal doses, it creates a slight increase in blood pressure, followed by a drop. In contrast, it does not seem to affect the rate of breathing.

Sexual Effects

Ginseng has long been reputed to have sexual effects. This is probably because ginseng dilates blood vessels, and blood flow to the genitals increases during sexual arousal. A study in humans showed that ginseng helped men with erectile dysfunction. Ginseng also seems to increase the sex hormone testosterone in men, which could also improve sexual function.

Hormonal Effects

Ginseng increases the release of stress hormones. Although this usually has a negative effect on mental functions, it is believed that over time the brain and body make adjustments and become less sensitive to these hormones. For example, when taken chronically ginseng reduces the amount of stress hormones that are released specifically during stressful events.

Ginseng also reduces the hormone prolactin, which is necessary for breast-feeding. As mentioned previously, it increases testosterone in males.

Cognitive Effects

Much research has been done on the effects of ginseng on memory. Unfortunately, most of it has been done in animals, with very little human research to match it. As far as animal studies are concerned, ginseng does improve many forms of memory. However, this does not mean it does so in humans and to an appreciable degree.

The few existing studies of ginseng's effects on human mental functions suggest a possible benefit. For example, one study showed that it caused improvement in attention and mental processing speed. Other studies suggest that a combination of ginkgo and ginseng improves memory function in healthy middle-aged people. What little evidence we have for ginseng is positive, but much more research is needed about its effects on human mental functions. No firm conclusions can be made at the present.

Side Effects

Studies of ginseng show generally low levels of toxicity. However, a "ginseng abuse syndrome" was reported in people taking very high doses (15 g/day, instead of the usual 1-2 g/day). With such high doses, people experience elevated blood pressure, irritability, nervousness, and sleeplessness. Usual doses of ginseng are not associated with serious adverse effects.

Caution has been suggested regarding a potential interaction between ginseng and warfarin. However, the interaction may be insignificant at normal doses.

TOBACCO

Few people think of tobacco *(Nicotiana tabacum)* as a drug that enhances mental abilities. It is more commonly known for its stimulant and addictive properties. Clearly, the health risks caused by tobacco use rule against it being used to improve memory in a therapeutic way. As mentioned in Chapter 4, nicotine is the main active drug in tobacco. It attaches to acetylcholine receptors and stimulates them. Since acetylcholine activates parts of the brain that are important for memory and other mental functions, it is not surprising that tobacco can improve memory.

Nicotine is a powerful drug, and taking it over a period of time causes the brain to adapt to it. So, when thinking about the effects of nicotine, it is important to consider whether users have taken the drug before, how much they take, and for how long they have been taking it.

Nicotine's Effects in Nonsmokers

Giving nicotine to people who do not use tobacco produces small but measurable improvements in mental functions. These include attention, speed of information processing, and memory.

The Effects of Abstinence in Smokers

Just as giving nicotine can improve mental functions, taking it over a long period of time and suddenly stopping can produce deficits. When cigarette smokers stop smoking, their mental functions decline somewhat, but if the nicotine is replaced (such as with the nicotine gum or patch), their functioning is the same as that of normal nonsmokers. For example, abstinent smokers have problems with short-term memory. These changes can occur as soon as twenty-four hours after quitting.

These impairments in mental functioning are part of the incentive to keep smoking. In addition to feeling bad, abstinent smokers have more trouble thinking and concentrating. Smoking again, however, relieves this problem. These changes are of practical daily importance. For example, driving an automobile depends on many complex mental skills such as reaction time and concentration. When abstinent smokers are given nicotine, their driving ability improves. Fortunately, replacing the lost nicotine with the patch or gum also helps. Quitting both tobacco and caffeine (e.g., in cola, coffee, tea, etc.) at the same time makes the situation even worse. Although it may be in a person's best interest to quit both drugs, it should be done in a more gradual way to make it tolerable.

The Effects of Abstinence in People with Alzheimer's Disease

Since nicotine stimulates the acetylcholine system in the brain, and Alzheimer's disease involves a loss of acetylcholine, could nicotine help people with Alzheimer's disease? The answer is actually yes.

When people with Alzheimer's disease are given nicotine, their memory improves somewhat. In fact, since a brain afflicted with Alzheimer's is deprived of acetylcholine, it is actually oversensitive to nicotine. Obviously the harmful effects of tobacco smoking make this impractical, but a nicotine patch is effective.

Conclusions About Nicotine

Nicotine has been shown to benefit attention and memory in animals and humans alike. This occurs in nonsmokers and even more strongly in abstinent smokers, since their brains have adapted to the presence of nicotine. Very few studies have been performed in Alzheimer's disease, but it seems to help there too.

Not only do the dangers of smoking (e.g., cancer, emphysema, heart disease) make nicotine an impractical drug, but other characteristics such as addiction and other side effects may completely rule out any therapeutic use of nicotine. Here modern pharmacology steps in: nicotine itself may not be useful, but it's possible that another similar drug can be made that has the same benefits but a better profile of side effects. For example, an experimental drug called ABT 418 works very similarly to nicotine but seems to have minimal side effects.

SOLANACEOUS VEGETABLES: (TOMATO, POTATO, AND EGGPLANT)

The possibility that common **solanaceous vegetables** such as tomato *(Lycopersicon esculentum),* potato *(Solanum tuberosum),* and eggplant *(Solanum melongena esculentum)* could affect the brain seems highly unlikely. However, these plants all belong to the Solanaceae family, and contain the chemicals alpha-solanine and alpha-chaconine. These chemicals prevent the breakdown of acetylcholine, causing it to accumulate. Most of the time, the effects are too small to be noticeable. However, outbreaks of potato poisoning have occurred. The greatest amounts of alpha-solanine and alpha-chaconine occur in potato sprouts, potatoes that have already sprouted, potato peelings, and potatoes greened by the sun. Although a normal meal of potatoes can provide 20 to 60 mg of alpha-solanine and alpha-chaconine, higher doses are needed to cause poisoning. The symptoms of potato poisoning are due to having too much acetylcholine, and include digestive disturbances such as stomachache, nausea, and vomiting.

Mental symptoms can include apathy, drowsiness, confusion, visual disturbances, dizziness, hallucinations, and trembling. The symptoms can persist from two to twenty-four hours after consumption.

However, these chemicals can cause problems at doses lower than those that cause full-blown potato poisoning. For example, even at normally consumed doses they can slow down the metabolism of drugs used for general anesthesia during surgery, causing a longer recuperation time.

The main purpose for mentioning these chemicals here is that their mode of action is identical to that of medications for Alzheimer's disease. It has not been formally researched, but it is possible that these two can interact, perhaps by boosting the effects of the medications. This is not necessarily a good thing: too much can increase the negative side effects of the medication. The only sound practical advice is that if a person is taking medications for Alzheimer's disease, it may be useful to pay close attention to the foods he or she eats. If side effects are experienced, and the person regularly eats potatoes, tomatoes, or eggplant, it may be helpful to keep a log to see if a relationship exists.

DISCUSSION AND TREATMENT OF PARKINSON'S DISEASE

Parkinson's disease is a neurological disorder characterized by problems with movement, thinking, and mood. It affects about 1 percent of the population over age fifty, and typically affects people aged forty to seventy years. However, it sometimes occurs at younger ages, as in the case of the well-known actor Michael J. Fox. The exact cause of Parkinson's disease is unknown, but it may involve a genetic vulnerability combined with exposure to environmental toxins. For example, lifetime exposure to certain pesticides, or a combination of pesticides, may be a factor. Welders have a greater risk of developing the symptoms of Parkinson's disease due to exposure to occupational toxins.

The motor symptoms of Parkinson's disease include difficulty initiating movement, slowed movement **(bradykinesia),** and muscle rigidity. One of the most commonly known motor symptoms is the tremor, which mainly occurs when the person's limbs are at rest. Depression is also common in people with Parkinson's disease, occurring in at least half of all cases. Of course, having a debilitating illness would be likely to make a person depressed, but something about the disease itself makes it more likely. Parkinson's disease also makes a person unable to show emotions by facial expression. Even though the person may be feeling the emotions internally, the disease reduces the ability to show them.

Parkinson's disease causes difficulties with mental abilities such as short-term memory, executive functions (e.g., planning, mental flexibility), and visual perception. Unlike Alzheimer's disease, most people with Parkinson's disease do not lose their long-term memories. Some cases, however (about 30 to 50 percent), develop Parkinson's disease plus an Alzheimer-like dementia, so that the person shows symptoms of both illnesses.

Although we don't yet know what causes Parkinson's disease, we do know its consequences in the brain. Parkinson's disease involves degeneration of a part of the brainstem called the **substantia nigra.** Neurons in this area supply other parts of the brain with dopamine, so when they start to die, the brain suffers a deficiency of dopamine.

Drug Treatment

Since Parkinson's disease results from a lack of dopamine, it was reasoned that replacing the dopamine might improve the symptoms. This treatment strategy is very effective, and adds years of function and independence to an afflicted person's life.

Dopamine replacement is accomplished with the drug **L-dopa,** the precursor for dopamine. Dopamine itself can't pass from the blood into the brain, so giving a person dopamine won't work. But L-dopa can pass into the brain, where it is converted into dopamine. The increase in dopamine in the brain temporarily reduces or eliminates the symptoms of Parkinson's disease. Another drug, **carbidopa,** is given along with L-dopa since it helps the L-dopa get into the brain. As a result, a lower dose of L-dopa is needed and fewer side effects occur (see Figure 5.3). These two medications (L-dopa and carbidopa) are sold in combined form under the brand name Sinemet.

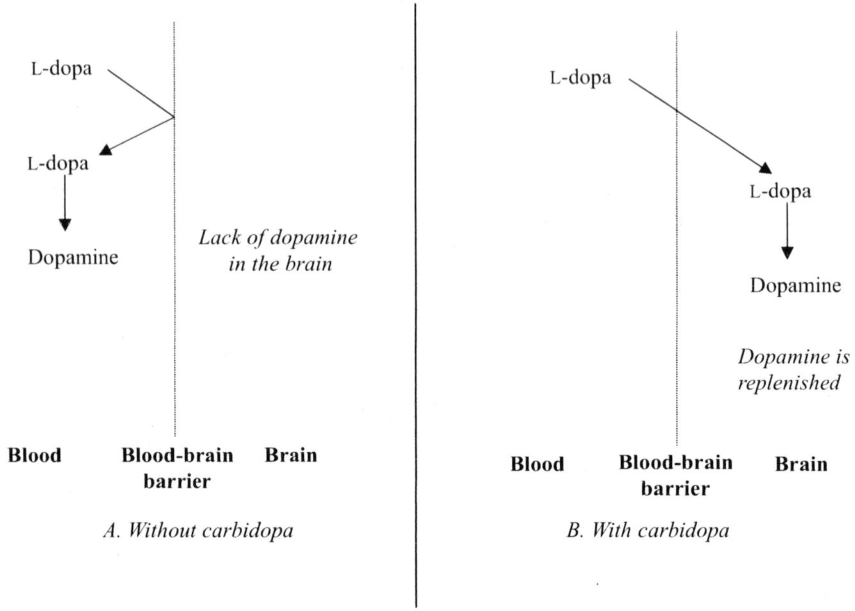

FIGURE 5.3. A. Without carbidopa, most of the L-dopa stays in the blood, so the brain remains deficient in dopamine. B. With carbidopa, more L-dopa gets into the brain, where it's converted to dopamine, replenishing the deficiency caused by Parkinson's disease.

Although very effective in most people, L-dopa is not a miracle cure. It loses effectiveness over time, producing fluctuations in symptoms, so higher doses are eventually needed. At higher doses, people develop excessive, uncontrolled, involuntary movements (called **dyskinesia**), which include facial grimaces, head wagging, writhing movements, and general restlessness. Also, the excess dopamine can cause delusional thinking, and/or hallucinations in 15 to 40 percent of people taking L-dopa. Although reducing the medication can reduce the hallucinations, it also worsens the motor symptoms, presenting a difficult situation.

Other medications for Parkinson's disease include drugs that do not increase dopamine, but rather directly stimulate the dopamine receptors (skipping dopamine altogether). One such drug is called bromocriptine, sold under the brand name Parlodel (see Figure 5.4).

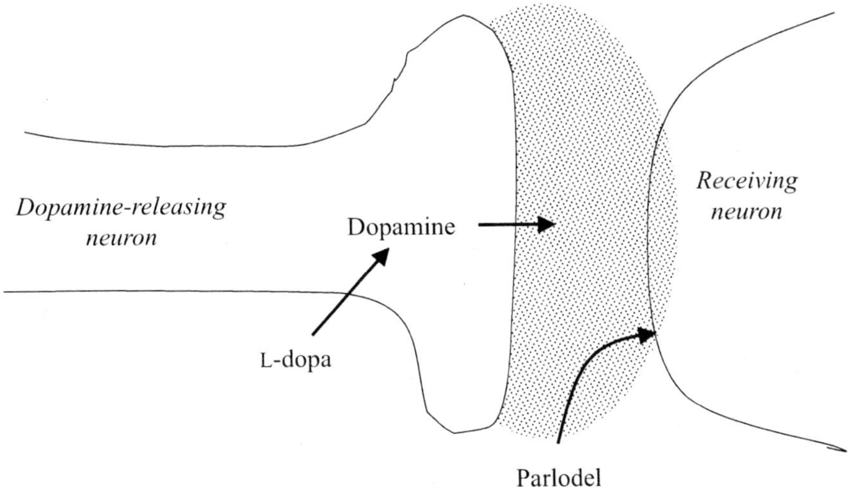

FIGURE 5.4. L-dopa is used by some neurons to produce the neurotransmitter dopamine. Giving a person L-dopa increases the amount of dopamine produced by the sending neuron. An alternate strategy bypasses the sending neuron and instead directly stimulates the receiving neuron. Thus, drugs such as Parlodel can work even after the sending neuron (left side of the diagram) dies, while L-dopa cannot.

Although L-dopa and Parlodel reduce the symptoms of the disease, they do not stop it from progressing. Another drug, Eldepryl, protects neurons from dying, at least to a limited degree. It is hoped that better drugs will be developed in the future to protect neurons from degeneration in Parkinson's disease.

FAVA BEANS

Fava beans *(Vicia faba),* also called broad beans, are commonly used in many cuisines. They are popular in Europe, Asia, the Middle East, Africa, and South America. *Fava* is the Italian word for the bean, which derives from the Latin word for the bean, *faba.* Anecdotal reports from people with Parkinson's disease suggested that the beans may have some effect on their symptoms. This possibility has been tested clinically, and it seems to be rooted in reality. Fava beans contain significant amounts of L-dopa.

A group of researchers had a group of people with Parkinson's disease stop taking their medications. After twelve hours off medications, they ate 250 g (about half a pound) of cooked fava beans. Keep in mind that most Parkinson's disease medications wear off before twelve hours, so by this time they were experiencing the full symptoms of their illness. However, after eating the fava beans, they noted significant improvement in their motor symptoms, equivalent to standard doses of pharmaceutical L-dopa. In fact, half of the people in the study with Parkinson's disease developed dyskinesias that occur when a person has too much L-dopa. As expected, blood levels of L-dopa increased in a manner comparable to taking pharmaceutical L-dopa. So, the fava beans were supplying the L-dopa.

In addition, five healthy volunteers ate fava beans and had a smaller increase of L-dopa in their blood. However, unlike the Parkinson's patients, the healthy volunteers did not develop the abnormal movements, probably because the Parkinson's patients had some carbidopa left over in their blood. Since carbidopa helps L-dopa get into the brain, the Parkinson's patients had large increases of L-dopa. In healthy people, however, most of the L-dopa stays in the blood and never gets to the brain.

Will fava beans be used as an alternative to pharmaceutical medications for Parkinson's disease? Probably not, for a few reasons. First, different batches of fava beans may have different concentrations of L-dopa, so the dosage cannot be controlled. Even with precise doses, it can be hard to control the symptoms of Parkinson's disease. Second, people with Parkinson's disease need L-dopa several times a day to keep steady levels in the blood and brain. It would mean eating fava beans several times a day, which would become tedious.

Since fava beans aren't likely to replace conventional medications for Parkinson's disease, they are mentioned here because of their potential interactions and side effects. Imagine what would happen if someone with Parkinson's disease took L-dopa and then ate fava beans! It would be equivalent to doubling the medication, and the person would likely experience the effects of overdose (e.g., excessive, abnormal movements).

SUMMARY

It is a common desire of healthy people to improve their mental functions and optimize their performance. In the case of people with dementia, cognitive-enhancing herbs offer the possibility of preserving mental functions or prolonging independence. The main strategies in this category include boosting neurotransmitters that are responsible for memory and thought (e.g., acetylcholine).

Ginkgo may be the best of this category, with the potential to boost neurotransmitters and preserve neurons. The present evidence suggests that it is probably as good as, but not better than, pharmaceutical treatments for Alzheimer's disease. It may help other conditions (e.g., vascular dementia and normal aging), and may even confer mental benefits on healthy people as well. The chief concern with ginkgo surrounds its inhibition of blood clotting, which may be dangerous to people taking blood thinners or who have forms of hemophilia.

Ginseng has good research in animal studies, but human studies are lacking. Tobacco works, but the long-term health risks and side effects (not to mention addiction) limit its usefulness.

Both solanaceous vegetables (tomato, potato, and eggplant) and fava beans may interact with pharmaceutical medications. Despite being commonly consumed in the diet, they may exacerbate the side effects of certain medications. Their therapeutic benefits have not been proven through research, and the prospect of consuming such foods on a daily and indefinite basis limits their potential therapeutic value.

Chapter 6

Sedative and Anti-Anxiety Herbs

Living in a complex and dangerous world requires us to respond when met with a challenge. The term *arousal* is used in different ways, often to mean increasing alertness or an emotional reaction. When faced with a predator, for example, an animal reacts physically and mentally to deal with it appropriately, either fighting or escaping. Although humans living in civilization may face such threats less often than our distant ancestors, the arousal machinery of the brain is still in place. Ironically, a physical and mental reaction that evolved to prepare us for an attacking lion is the same response elicited by a college exam. Still, the arousal reaction is appropriate in some situations. But as with many disorders of the mind and brain, a normal and adaptive reaction can be overactivated and become maladaptive. Overactivation of our arousal mechanisms can result in anxiety and insomnia.

THE PROBLEM OF ANXIETY

Anxiety is difficult to capture in words, but it's one of the basic, universal emotions experienced by most humans. It has many components, including emotional, cognitive, physiological, and behavioral aspects. Emotionally, anxiety is described as feelings of unease, discomfort, or dread. It is most often an unpleasant state and reduces our capacity for experiencing pleasure. The cognitive component consists of the thoughts that occur in anxiety, such as worrying and pessimistic thinking. The physiological side is all too familiar: muscle tension, rapid heart rate, raised blood pressure, sweating, dry mouth, nausea, dizziness, and diarrhea. This unpleasant state causes us to avoid the source of anxiety.

Low levels of anxiety are helpful, keeping us alert and appropriately concerned. Some neurological and psychiatric disorders cause a loss of anxiety, and the results are detrimental. Such people fail to heed warnings and lack concern in respect to danger. On the other end of the spectrum, anxiety is just as bad when it becomes too severe, frequent, or lasts too long, since it interferes with our ability to function. For example, constant anxious thoughts hinder the ability to concentrate.

Anxiety disorders are a diverse group that includes panic disorder, phobias, obsessive-compulsive disorder, PTSD, substance-induced anxiety disorder, and generalized anxiety disorder. Generalized anxiety disorder means that a person experiences persistent, diffuse anxiety. The exact source of the anxiety is unknown, and the feeling pervades many aspects of life. A substance-induced anxiety disorder occurs in reaction to taking or stopping a drug. This includes, for example, drinking too much caffeine or suddenly ceasing to drink alcohol after a long period of use. A phobia is an intense, excessive anxiety reaction to a specific situation or object. However, the fear must be entirely out of proportion to the actual threat posed. For example, social phobia means that a person is overly concerned about embarrassment or being judged negatively in social situations. Panic disorder means that one experiences sudden, intense waves of anxiety and dread. They may come out of nowhere and rapidly peak. Sometimes anxiety from a phobia can reach panic-level intensity (e.g., getting stuck in an elevator when one is afraid of closed spaces). PTSD is a condition that sometimes follows a severe stressor or trauma, such as combat or a robbery. Its symptoms include recurrent and intrusive memories of the trauma and feelings of detachment. People with obsessive-compulsive disorder experience obsessions (repetitive, intrusive thoughts) and compulsions (repeated, ritualized behaviors that reduce the anxiety caused by the obsessions). Despite their differences, all of these disorders share anxiety as a principal symptom, of greater duration or intensity than normal experience and leading to difficulty functioning in everyday life.

INSOMNIA

Insomnia is a persistent problem with sleeping. It may be a reduction in the amount of time sleeping, or poor-quality sleep, when one wakes up and doesn't feel refreshed or rested. Insomnia can be the primary problem that a person experiences, or it can be part of other problems. For example, anxiety can cause insomnia, because a person may have trouble falling asleep due to repetitive thoughts and worrying at night. Depression can also cause insomnia, but the pattern is typically different. Depressed people may wake up very early in the morning and may be unable to return to sleep, even though they feel tired. Sleep apnea is another disorder that ruins the quality of sleep. In this case, breathing is obstructed and the person is frequently aroused throughout the course of the night. This prevents deep, restful sleep even though the person may spend an entire night without fully waking up.

THE STRUCTURE OF SLEEP

Sleep and its disorders are most often studied by using the EEG. With this instrument, electrodes are painlessly attached to the scalp, and patterns of the brain's electrical activity are measured throughout the night. When a person is awake and active, the brain shows beta activity (see Table 6.1), which has the fastest and most asynchronous brain waves, small and choppy like the waves in a pool full of children. When a person is awake but quiet and resting, the brain shows more alpha activity, which is a little bit slower. Upon falling asleep, the brain starts to show theta waves, which are even slower. Finally, after some time, the brain shows delta activity, which has the slowest and biggest waves. (Imagine the waves from one person in a pool with an inner tube, slowly bobbing up and down.) Delta activity occurs when we are in the deepest, most restful periods of sleep.

TABLE 6.1. Brain wave activity and behavioral states.

State	Type	Frequency		Synchrony
Awake, alert	Beta	15-35 Hz	(fastest)	Least synchronous (choppy)
Awake, resting	Alpha	8-12 Hz	↑	
Sleep	Theta	4-7 Hz	↓	
Deepest sleep	Delta	1-3 Hz	(slowest)	Most synchronous (smooth)

Brain activity fluctuates in cycles throughout the night, with periods of greater or lesser activity. At some points, we enter **REM sleep,** named after the rapid eye movements that occur. Ironically, the brain shows beta waves, representing high levels of activity, almost as if it were awake. This is the period when the most vivid and life-like dreams occur. Unfortunately it is not a very restful state because the brain is so active. Medications that are prescribed for sleep disorders tend to increase slow-wave stages of sleep, namely theta and delta activity.

CENTRAL NERVOUS SYSTEM DEPRESSANTS

Medications that slow down activity in the brain are called central nervous system (CNS) depressants. The word *depressant,* however, should not be confused with emotional depression. It simply means reduced activity in neurons.

Many CNS depressants are used to treat anxiety, epilepsy, and insomnia. They are also used to induce unconsciousness for major surgery (general anesthetics), and are used for recreation (e.g., alcohol) (see Table 6.2). All CNS depressants produce a similar range of ef-

TABLE 6.2. Examples of CNS depressants.

Class	Generic name	Trade name
Barbiturate	Phenobarbital	Luminal
	Pentobarbital	Nembutal
Benzodiazepine	Alprazolam	Xanax
	Diazepam	Valium
	Lorazepam	Ativan
Alcohol	Ethanol	
Antiseizure	Carbamazepine	Tegretol
	Gabapentin	Neurontin
	Phenytoin	Dilantin
	Tiagabine	Gabatril
General anesthetic	Chloroform	
	Halothane	Fluothane

fects, depending on their dosage and potency. At small doses, they tend to relieve anxiety (see Figure 6.1). With larger doses, a person sheds some inhibitions, becoming more sociable or behaving foolishly. At even larger doses the person becomes sleepy and may lose consciousness. High enough doses cause general anesthesia, so that intense pain won't even wake up the person. Eventually, coma and death result because breathing is suppressed.

Different drugs work best in different parts of the range of CNS depressants. For example, **benzodiazepines** (e.g., Xanax, Valium) work best for reducing anxiety and inducing sleep. However, **barbiturates** (e.g., phenobarbital, pentobarbital) cause too much sleepiness to be useful in treating anxiety. Halothane is used for general anesthesia, and would be practically useless in treating anxiety.

The loss of inhibition that can occur from CNS depressants (such as alcohol) may seem paradoxical. Instead of being sedated, a person may become more animated and active at first. This occurs only at lower doses, because specific areas of the brain are suppressed. Parts of the frontal lobe that control and inhibit our behavior are very sensitive, so they are affected first before the full sedative effects kick in.

Issues with CNS Depressants

A few issues pertain to most CNS depressant drugs. When two or more CNS depressants are combined, they can produce additive or synergistic effects (multiplying each other's effects, rather than simply adding). Taking a sedative drug at a certain dose may be perfectly safe, but when combined with other sedatives (e.g., Xanax and beer), the effects can be unexpectedly extreme, and sometimes lethal.

Many CNS depressants have some potential to cause dependence. The barbiturates have more abuse potential than the benzodiazepines.

Lowest doses	Reduction of anxiety
↑	Reduced social inhibitions
	Sedation
	Sleep (can be awakened by pain)
	General anesthesia (cannot be awakened by pain)
↓	Coma
Highest doses	Death (stops breathing)

FIGURE 6.1. Continuum of effects of CNS depressants.

Many antiseizure medications do not seem to cause dependence, at least not in the sense of addiction. However, CNS depressants are known to produce tolerance when taken over time. Increasing doses are needed to maintain the same level of effect. **Cross-tolerance** also occurs, meaning that developing a tolerance to one CNS depressant automatically causes tolerance to other similar drugs. As tolerance increases, the safety margin gets narrower: the difference between a dose and an overdose becomes smaller.

People with tolerance to CNS depressants experience a withdrawal syndrome if the drug is suddenly stopped. Persistent suppression of activity in the nervous system by the drug causes it to adapt in an attempt to compensate. This is called a rebound effect: the brain becomes more sensitive, and if the drug is suddenly taken away, the brain is then oversensitive and overactive until it has time to return to normal. The degree to which this happens depends on how much drug is taken and for how long. For mild cases it may only cause anxiety and irritability. More severe cases involve delirium (e.g., delirium tremens from alcohol), confusion, and seizures.

To avoid this, the drug must be withdrawn slowly, giving the brain time to adapt gradually. Sudden withdrawal from a CNS depressant, such as alcohol, is dangerous and can present a medical emergency. In the case of alcohol, people are switched to a safer drug such as a benzodiazepine and withdrawn gradually over weeks, which minimizes the withdrawal symptoms.

How CNS Depressants Work

Neurons can either excite each other or inhibit each other. So, CNS depressant drugs can either prevent that excitation or increase inhibition, which is exactly how they work. This can happen in several ways.

Many CNS depressants increase the actions of the neurotransmitter GABA. GABA has the effect of calming activity in the nervous system, so CNS depressants augment this effect. Benzodiazepines, barbiturates, and alcohol work this way, as do some seizure medications.

Other CNS depressants block the excitation of neurons. Glutamate is a neurotransmitter that strongly activates neurons, and it is blocked by drugs such as alcohol, PCP (known as angel dust), and ketamine (known as Special K). Alcohol has a double sedative action, blocking excitation and increasing inhibition. Glutamate is very important for making memories, so it is not surprising that drugs such as alcohol cause memory impairment when doses are high enough.

CNS depressants can also reduce the excitability of neurons by blocking sodium (Na^+) or calcium (Ca^{2+}) channels in neurons. This makes the neurons less electrically excitable and also has sedative effects. Some of the drugs that do this are used as local anesthetics, since they block electrical signals in pain nerves.

VALERIAN

Valerian grows in most parts of the world, but is native to Europe and Asia. There are many similar species, but the best known is *Valeriana officinalis* (see Table 6.3). It is a flowering herb with an erect stem and many small pink-white flowers.

Valerian has been used for more than 1,000 years, and was known in sixteenth-century Europe as a treatment for epilepsy. Among its reported uses are treatment of insomnia and anxiety. Anyone familiar with valerian knows its distinct, unpleasant odor.

Active Chemicals in Valerian

Valerian contains many chemicals, and it is not certain which one or ones are responsible for the sedative effect. Some of the candidates

TABLE 6.3. Herbal CNS depressants.

Herb	Active chemical	How it works
Catnip	*Cis-trans*-nepetalactone	Unknown
Chamomile	Apigenin	Benzodiazepine receptor
Hops	2-methyl-3-butene-2-ol	Unknown
Kava	Kavalactones	Increases GABA Blocks sodium and calcium channels Increases norepinephrine and dopamine
Passionflower	Chrysin	Benzodiazepine receptor
Valerian	Sequiterpenes	Prolongs GABA effects

include isovaleric acid, valerenic acid, valerenal, valeranone, and valerenol.

How Valerian Works

Exactly how valerian works is uncertain, but there is reason to believe that it works through GABA. Chemicals found in valerian seem to increase the activity of GABA, which would calm the activity of neurons. More research is needed to have a better understanding of how valerian works.

The Effects of Valerian

Muscle Relaxant Effects

Valerian seems to have a muscle relaxant effect by directly affecting muscles. However, because GABA is involved in relaxing muscles through the spinal cord, valerian could also create these effects through GABA.

Sedative Effects

Animals given valerian show sedation. They become less active, and valerian reduces hyperactivity caused by the administration of caffeine. Compared to other CNS depressant herbs, valerian is more of a sedative, whereas passionflower and chamomile can have anxiety-reducing effects without sedation at certain doses.

Antiseizure Effects

Valerian extracts have mild antiseizure effects in mice. This is consistent with the possible effects on GABA, since other antiseizure medications influence GABA. The historical notion that valerian reduces seizures may be vindicated. However, it has never been tested in humans. Valerian may have been the best choice for treating seizures when no alternatives were available. It may be ineffective compared to modern pharmaceutical antiseizure medications. Valerian should not be used in place of seizure medications to treat epilepsy.

Further, combining valerian with seizure medications may produce oversedation and could be dangerous.

Valerian and Sleep

Some good evidence suggests that valerian improves sleep. A sleep study using EEG showed that valerian lengthened certain stages of sleep. The effects were also dose dependent: a larger dose produced greater sedative effects. The maximum sedative effect occurred two to three hours after the valerian was taken. Another study showed that valerian increased deep, slow-wave sleep in people who were poor sleepers.

Whereas these studies examined the effects of valerian on brain activity in sleep, other controlled studies analyzed the subjective effects. Compared to placebo, valerian improves the quality of sleep and shortens the amount of time it takes to fall asleep. Further, the improvement in sleep was most noticeable in poor sleepers and tobacco smokers.

More research is needed to truly verify the benefits of valerian in sleep. However, these studies offer some proof that valerian works.

Cognitive Effects

Very few studies exist on the cognitive effects of valerian. One study compared a combination of valerian and hops against the sleep medication Rohypnol (flunitrazepam), which is a benzodiazepine. Rohypnol caused hangover effects, making thinking sluggish, but valerian and hops did not. About one to two hours after taking it, valerian does slightly reduce a person's ability to pay attention. It would be safer to avoid taking valerian during activities that require sharp attention, such as driving, especially at night.

Side Effects

Raw valerian contains chemicals called valepotriates, which could cause liver damage. However, the amount of these chemicals in *Valeriana officinalis* is very low, and they are not well absorbed into the body. Further, they rapidly break down when the herb is dried and stored. No occurrences have been reported of liver toxicity in animals or humans due to valerian. In fact, no health hazards have been reported with proper use of valerian, and it has been approved by the German Commission E as a treatment for anxiety and poor sleep.

Overdoses of valerian cause headache, vomiting, stupor, dizziness, and heart irregularities. One case has been reported of a person who took a single overdose (approximately twenty times the recommended dose), which produced relatively mild symptoms and resolved in twenty-four hours. Since valerian has CNS depressant effects and may work through GABA, it should not be combined with other sedative drugs.

Summary of Valerian

Good preliminary evidence shows that valerian is useful for improving sleep. It's not certain which chemicals in valerian are responsible, but they may work through enhancing the neurotransmitter GABA. Whether or not valerian is useful in treating anxiety has not been tested. Side effects appear to be minimal with proper use.

KAVA

The **kava** plant *(Piper methysticum)* is native to many of the South Pacific islands, where it is known by many different names (see Figure 6.2). South Pacific cultures have used kava as a ceremonial or rec-

FIGURE 6.2. Kava *(Piper methysticum).* Copyright 2004 by Marcello Spinella.

reational herb. It was often given as a gift to resolve social conflicts, or partaken as an after-work drink to relax. It is known traditionally for producing relaxation. Europeans discovered it when they traveled to the South Pacific islands in the eighteenth century. Unlike many other sedative herbs and pharmaceutical drugs that cause sleepiness, kava is reputed to produce a relaxed but alert mental state. It can be taken for sleep in higher doses.

Kava is a member of the pepper family and its botanical name, *Piper methysticum,* means "intoxicating pepper." Kava grows as a large bush two to three meters in height, with large heart-shaped leaves. The part of the plant used medicinally is the rhizome, or underground stem, which is thick and woody. It was traditionally prepared by pounding, chewing, or grating, and then soaking it in cold water, which was then drunk. In the West, it is typically sold dried and ground in capsules, either whole herb or extract. The extract is the best preparation to use. The teas may be less effective.

Active Chemicals in Kava

The active chemicals in kava are called the **kavalactones.** These include kavain, dihydrokavain, yangonin, dimethoxyyangonin, methysticin, and dihydromethysticin. These chemicals are known to cross the blood-brain barrier and enter the brain.

How Kava Works

Kava seems to create its effects by a combination of actions. It enhances the neurotransmitter GABA. The benzodiazepine drugs, commonly prescribed for anxiety and sleep, work through a similar, but not identical, mechanism. Kava also reduces the electrical excitability of neurons by blocking sodium and calcium channels. These combined effects produce significant CNS depressant effects.

Kava also may increase the effects of the neurotransmitters norepinephrine and dopamine in certain areas of the brain. This effect is uncommon to other sedative drugs. In fact, increases in norepinephrine and dopamine are more likely to produce alertness. This may explain why kava produces a calm but alert state at lower doses.

Drug Issues with Kava

For a drug to have psychoactive effects, it must gain entry to the brain, and kavalactones have been shown to do so. Kavalactones also are more effective when taken in combination, rather than individually. It's not certain why, but taking them in combination causes a greater proportion of them to enter the brain. This occurs normally when someone consumes the whole herb or extract. Isolating single chemicals from kava would probably not be useful, whereas it may be useful in the case of other herbs.

How kava is prepared alters its effectiveness. Some of the chemicals in kava are fat soluble and less soluble in water. Water extracts tend to be less effective than those which have the fat-soluble chemicals.

Tolerance seems to be limited in kava. When kava was given daily to mice, its effects did not seem to decrease. However, combining kava and ethanol or barbiturates produced strong sedative effects.

The Effects of Kava

Muscle Relaxant Effects

The muscle relaxant effects of kava have been studied only in mice and frog muscles. However, its effects probably pertain to humans as well.

Analgesic Effects

Kava has also been shown to have pain-killing effects. This has been researched only in animals, but ounce for ounce, kava seemed to be as effective as aspirin. Studies in humans are needed to confirm this effect.

Sleep

The effects of kava have not been formally researched on sleeping humans. However, EEG studies show that kava increases slow-wave brain activity, which would improve deep, restful sleep. Kava has a traditional reputation among Pacific Islanders for promoting sleep. It would be nice to see some scientific studies to complement the traditional knowledge in this respect.

Cognitive Effects

When taken alone and in low doses, kava seems to have minimal effects on mental abilities. Kava has very small effects on tests of mental abilities, such as attention, reaction time, and short-term memory. However, when combined with alcohol, kava produced more significant effects.

Studies have examined the effects of kava on brain electrical activity during thinking tasks, such as memory and visual attention. They also compared kava to the anti-anxiety drug Serax (oxazepam), which is a benzodiazepine drug. Serax caused cognitive impairments that were reflected in the electrical measures. However, kava did not.

So even though kava produces relaxation, it has minimal effects on mental abilities compared to an anti-anxiety drug and alcohol. This is a useful feature, since most anti-anxiety drugs produce unwanted mental impairments.

Antiseizure Effects

Kava has some antiseizure effects on certain types of seizures in animals. As with other herbal sedatives, it is not clear whether it works in humans. It should not be used as a replacement for prescribed antiseizure medications. Also, since its effects are similar to those of some antiseizure drugs, such drugs should not be taken in combination with kava.

Anxiolytic Effects

There is good evidence for the anti-anxiety effect of kava. Studies have used doses of 60 to 240 mg of kavalactones per day, and lasted between four and twenty-four weeks. Kava has been shown to be superior to placebo for reducing anxiety. One study showed that kava was as effective as the prescription anti-anxiety drug Serax (oxazepam).

Side Effects of Kava

No health hazards are known with the proper use of kava. Kava has been approved by the German Commission E for treatment of anxiety and insomnia. In clinical studies of kava for anxiety, adverse effects were uncommon and did not differ across placebo and kava groups.

A case has been reported of a man taking 210 to 280 mg of kavalactones per day for three to four months who developed hepatitis. This is a relatively large dose to take on a regular basis. To be safe, high doses and prolonged constant use of kava should be avoided until this issue is better understood.

Since kava is a sedative herbal medicine, it should not be combined indiscriminately with other sedatives such as benzodiazepines, barbiturates, ethanol, or antiseizure medications. One case has been reported of a man who combined kava with the anti-anxiety drug Xanax (alprazolam). He became disoriented and lethargic and returned to normal only after several hours.

Liver Toxicity

In several reports, people who took kava experienced liver damage. In such cases, it is always difficult to know whether kava caused the damage. In some cases, there was pre-existing liver disease that may have been exacerbated by kava. In other cases, people were taking high doses or taking kava in combination with other drugs that can harm the liver. In defense of kava, it has been used by Pacific Islanders for numerous generations, and that population does not appear to have a higher rate of liver disease. Nonetheless, to be safe, kava should not be taken in large doses, for prolonged periods of time, and especially not in combination with other drugs that can affect the liver (e.g., alcohol, Tylenol). Only continued research and vigilance in this area will give us the final answer on kava's safety.

Kava Dermopathy

A scaly skin condition called kava dermopathy has been reported as a result of heavy, long-term kava use. It is reversible, but has never been reported in the Western countries. It's possible that a vitamin deficiency, in addition to heavy kava use, is needed to develop this condition.

Parkinson's Disease

People with Parkinson's disease should avoid kava, or at least try it under the supervision of a physician. A few cases were reported of abnormal movements occurring in people with Parkinson's disease who took kava. The abnormal movements ceased after the kava was discontinued.

Summary of Kava

Kava has a traditional reputation among Pacific Islanders for causing a calm but alert state. The active chemicals seem to be the kavalactones, which create the sedative effects through the neurotransmitter GABA and by reducing electrical excitability in neurons. Although it does effectively reduce anxiety, it seems to have minimal effects on mental functioning. However, when combined with other sedatives, the effects may become more prominent and dangerous.

PASSIONFLOWER

Passionflowers are among the most beautiful and ornate species of flowers. A few members of the passionflower family have psychological effects. The three that have been studied are *Passiflora incarnata, P. coerulea,* and *P. edulis*. The fruit of *P. incarnata* is commonly known as passion fruit, which is sweet and is used to flavor many popular drinks. Native Americans have used passionflower as a tea for its sedative and anti-anxiety effects.

Active Chemicals in Passionflower

The active chemical in passionflower is not certain. However, the best candidate is a chemical called **chrysin.** It belongs to a family of similar chemicals called flavonoids.

Mechanisms of Action

Chrysin works at benzodiazepine receptors in the brain. These are the target sites for many common pharmaceutical anxiety and sleep medications such as Xanax, Valium, and Rohypnol. Further, it only partially activates these receptors, not reaching the full effect.

Effects of Passionflower

Sedative and Anti-Anxiety Effects

Since chrysin works in a manner similar to pharmaceutical anti-anxiety drugs, it is not surprising that passionflower has been associated with anti-anxiety effects. In fact, many similar flavonoids have been shown to have this effect.

As far as controlled, scientific research goes, however, passionflower has not been tested in humans. Thus, we don't know if it is really effective in humans, or how potent it is. Animal studies of passionflower do show anti-anxiety and sedative effects. Conveniently, this depends on the dose given (smaller doses reduce anxiety; larger doses cause sedation), so anxiety could be reduced without causing too much sleepiness. As with other herbal sedatives, combining it with another sedative drug can magnify its effects.

Cognitive Effects

Passionflower's effects on mental abilities have not been studied in humans. However, a study in rats found that chrysin did not impair memory, which is a major concern with most pharmaceutical benzodiazepine medications. Although human studies need to be done, this suggests it may be able to reduce anxiety without clouding thinking and memory.

Antiseizure Effects

Chrysin has antiseizure effects in mice. However, this effect has never been examined in humans, and passionflower should not be relied upon to treat a seizure disorder.

Toxicity

No adverse effects have been reported to result from proper use of passionflower. Like other herbs in this category, combination with other sedative drugs will magnify its effects.

CHAMOMILE

Chamomile has been used throughout history, including ancient Egyptian, Roman, and Greek cultures. The name *chamomile* refers to two similar species of plants: German chamomile *(Matricaria recutita)* and Roman chamomile *(Chamaemelum nobile)* (see Figure 6.3). Both look very much like daisies, with white petals and a yellow central disc. They are both natives of Europe, Africa, and Asia, and also

FIGURE 6.3. Chamomile *(Matricaria recutita).* Reprinted from Culbreth DMR (1927). *Materia Medica and Pharmacognosy,* Seventh Edition. Philadelphia: Lea & Febiger.

grow in North America. The most common means of taking them are to dry the flowering tops and use them to brew a tea. There are many reported uses of chamomile, but it is best known for mild calming effects and reducing muscle spasms.

Active Chemicals in Chamomile

Many chemicals may be active in chamomile. The best candidate for its sedative effects is **apigenin,** which belongs to the flavonoid family of chemicals.

How Chamomile Works

How chamomile causes sedation and calming is not entirely certain, but apigenin works at benzodiazepine receptors. As with passionflower, this is an effect common to pharmaceutical sedatives and antianxiety drugs, the benzodiazepines.

Effects of Chamomile

Chamomile's sedative and anti-anxiety effects have not been formally tested in humans. However, animal studies show anti-anxiety effects, and sedative effects at much larger doses. Chamomile does not appear to have any antiseizure or muscle relaxant effects. Chamomile does not seem to impair memory, even at doses higher than needed to reduce anxiety.

Side Effects

Chamomile appears to have very few negative side effects. The U.S. Food and Drug Administration has placed it on the Generally Regarded As Safe (GRAS) list. Some may have allergic reactions to the pollen in the flowers, but this is rare.

OTHER REPUTED SEDATIVE HERBS

A few other herbal medicines are reputed to have sedative effects: skullcap, geranium, hops, and catnip. However, scientific evidence for them has not yet been gathered.

Catnip

Catnip *(Nepeta cataria)* is a green-gray-leafed perennial plant with spotted white tubular flowers. Although better known for its effects in cats, it is known for its sedative effects in humans. In one case, accidental ingestion by a young child reportedly produced sedative effects. The active chemical is uncertain, but it may be a group called the nepetalactones. Cats have a very different reaction to catnip than humans. Nepetalactones are believed to mimic a scent in cat urine that triggers sexual behaviors.

Hallucinogenic effects of catnip have been erroneously reported in humans. It was later shown that this resulted from confusion of the plant with cannabis.

Hops

Hops *(Humulus lupulus)* is a flowering vine that grows in Europe, Western Asia, and North America. It is an ingredient of beer, but is also consumed as a tea that has calming and sedative effects. The female flowers are also placed in a pillow to treat insomnia. The chemical responsible for this may be 2-methyl-3-butene-2-ol, which causes sedation when given to mice.

Skullcap

Skullcap *(Scutellaria lateriflora)* has been used in Chinese and Western medicine for sedative and antiseizure effects. However, its effects have not been established in either animals or humans.

Lemon Balm

Lemon balm *(Melissa officinalis)* has sedative effects in mice, which have not been tested in humans. It also magnifies the effects of other sedative drugs.

SUMMARY

Several herbs have sedative and anti-anxiety effects. They all seem to work by reducing activity in the central nervous system, just as their pharmaceutical counterparts do. Evidence for their effects is based on neurochemistry and on their behavioral effects in animals and humans, although some have more proof than others. In some cases, such as kava, the evidence is fairly solid. In other cases, such as chamomile, there is good preliminary evidence, but more research needs to be done. For certain herbs, such as lemon balm and skullcap, many basic questions are unanswered, such as whether or not they really work.

Chapter 7

Herbs for Mental Illnesses

Several herbal medications have been used historically to relieve the symptoms of mental disorders. Drug treatment of mental illnesses really didn't become effective on a wide scale until the twentieth century, and many of the historical uses of plant drugs for this purpose are considered impractical by today's standards. For example, Sigmund Freud, the creator of psychoanalysis, advocated the use of cocaine to alleviate depression. Cannabis and opium were also used historically to treat agitation.

However, a few herbal medicines may fit the bill by today's standards. Many of them are still being tested for their effectiveness and safety, so they must be considered on a case-by-case basis. Some are explained here primarily for their historical interest.

THE ERA OF PSYCHOTHERAPEUTIC DRUGS

During the twentieth century, revolutionary discoveries of drugs for mental illness were made. Before those discoveries, psychotherapy was the only available option, and other known drug treatments were essentially ineffective. For example, drugs such as bromides and barbiturates were used to treat people with schizophrenia. Although the drugs sedated them, they did nothing to help their delusions or hallucinations, which are at the core of the disease.

The first modern drug treatment for a mental illness came from an Australian psychiatrist named John F. J. Cade. He proposed the use of lithium to treat bipolar disorder (manic-depressive mood swings), which was slow to gain acceptance but is now a standard treatment. In 1950, the drug chlorpromazine (sold under the brand name Thorazine) was made in France and found to be effective for treating schizophrenia. The first antidepressant drug, iproniazid, was tested in the late 1950s. Meprobamate (sold as Equanil or Miltown) was introduced as an anti-anxiety drug in the mid-1950s. Although it's no longer used, it sparked the discovery of a series of better sedative and anti-anxiety drugs.

Ironically, many of these drugs were discovered serendipitously. For example, chlorpromazine was originally developed as an antihistamine drug. Iproniazid was originally developed to treat tuberculosis. It was only by accident, and astute observation, that these drugs came to be used to treat mental illness.

The discovery of medications for mental illness was revolutionary because they provided a means to treat illnesses that were otherwise intractable. With the exception of electroconvulsive (or electroshock) therapy for severe depression, there were no medical treatments for mental illnesses that could not be cured by psychotherapy.

Another significant point about these medications is that they changed the way we think about mental illnesses. Indeed, the phrase *mental illnesses* reveals our conceptualization of them as diseases of the mind, but they are equally diseases of the brain. Because medications affect some of these illnesses, it supports the idea that they have a basis in the brain. The fact that these drugs worked at all led researchers to discover how the drugs were working and to speculate about what might be going wrong in the brain in the first place. The logical next step was to design new drugs based on more advanced understanding, rather than relying on pure chance.

Ideally, treatment of mental illnesses should involve both medications and therapy. Medications are no replacement for people learning new skills and ways of coping with their condition in everyday life. However, medications can improve the outcome of therapy, particularly in more severe cases. The two can be used together in a complementary manner.

The development of medications for mental illness has an ironic downside. Before these medications, many people were institutionalized in psychiatric hospitals, living as permanent or frequent residents. Some 550,000 patients were living in psychiatric hospitals after World War II, which was roughly 8 out of 2,000 people in the United States at the time. By 1986, this number had declined to 111,000, which is roughly one person for every 2,000. On the surface, that sounds like progress: the proportion of people who need to be hospitalized decreased. In reality, the people who improved somewhat were released with medications, but follow-up was poor. Schizophrenic patients, as well as others, are notoriously noncompliant in taking medications. As a result, many of them got worse after being released, were unable to live independently, and ended up living on the streets. Proper follow-up and community care are necessary for chronic psychiatric patients.

Shortcomings aside, medications for mental illness have alleviated an unimaginable amount of human suffering. With careful research, some herbal medications may add to this effort.

DEPRESSION AND ANXIETY

Depression is one of the most common mental disorders. It is a significant public health problem: at least 17 percent of people will be diagnosed with clinical depression at some point in life, and at any given time, approximately 5 percent of the population suffers from it. These statistics only include depression of clinical severity (known officially as major depression), but many more people suffer from milder forms that don't meet the full criteria for an official diagnosis. Sometimes, psychologists use the term *dysphoria* to refer to depression that is less severe.

Depression is characterized by intense sadness and impairment of thinking in the form of concentration problems, worrying, pessimism, and lowered self-esteem. Relationships suffer with depression and people tend to withdraw socially. This is ironically unfortunate, since social support is so important in recovery and prevention of depression. The brain and body undergo physiological changes in depression as well: appetite, sleep, sex drive, and hormonal levels are changed. Appetite deviates from the normal pattern, but it can either increase or decrease. Sleep is of poor quality and the person may wake early in the morning feeling unrested but unable to return to sleep. Stress hormones are overactive, and brain cells suffer damage over time as a result. Brain scans of people with chronic, recurring depression show atrophy in brain areas important for memory (i.e., the hippocampus).

Although sadness and grief are normal emotions in some situations, depression is a pathological condition distinguished by duration and severity. Sadness and grief are adaptive when they make people rest and take stock after a loss, and make adjustments in their lives in order to move on. On the other hand, depression lacks these adaptive advantages, making a person suffer too much and too long, without benefit.

Anxiety disorders are a mixed bag that all share anxiety as the core symptom. They are discussed in greater detail in Chapter 6. They are mentioned here because some of the herbs in this chapter may alleviate anxiety but are not considered sedatives.

Pharmaceutical Antidepressant and Anti-Anxiety Medications

Many antidepressant medications are on the market (see Table 7.1). These drugs have come in waves or generations. The first generation was the **MAO inhibitors,** which were followed by the tricyclics, and then the **SSRIs** (selective serotonin reuptake inhibitors). The latest drugs are diverse and not easily grouped, and sometimes are given the mixed-bag name "atypical" antidepressants. Each generation has been as effective as prior generations. However, the improvement is in the form of smaller side effects—a critical factor for com-

TABLE 7.1. Examples of pharmaceutical antidepressant drugs.

Type	Generic name	Trade name	Main drawbacks
MAOI	Phenelzine	Nardil	Dietary restrictions
	Tranylcypromine	Parnate	
Tricyclic	Imipramine	Tofranil	Blurred vision, memory problems, dry mouth, slight risk of abnormal heart rhythms
	Amitriptyline	Elavil	
	Clomipramine	Anafranil	
SSRI	Fluoxetine	Prozac	Sexual dysfunction, other effects
	Sertraline	Zoloft	
	Paroxetine	Paxil	
	Fluvoxamine	Luvox	
	Citalopram	Celexa	
Atypical	Nefazodone	Serzone	Other side effects
	Mirtazapine	Remeron	
	Buproprion	Wellbutrin	
	Venlafaxine	Effexor	

pliance. An effective medication only helps if people can tolerate taking it.

Despite all their differences, all antidepressant drugs seem to have one thing in common: they work on the neurotransmitters serotonin, norepinephrine, or dopamine. The MAO inhibitors block the enzyme that breaks the neurotransmitters down, prolonging their action. The tricyclics and SSRIs both block the neurotransmitter from being pumped back into neurons, extending the duration of their effects. (As the name suggests, the SSRIs do this selectively with serotonin.)

A problem with the MAO inhibitors is their side effects. These drugs also prevent breakdown of a chemical called tyramine, which is found in many foods such as aged cheese, red wine, and figs. If people take an MAO inhibitor with these foods, their blood pressure will rise, which could cause a stroke. So people taking those medications have to watch their diet carefully. The tricyclics do not have that dietary restriction, but they cause side effects such as dry mouth, blurred vision, and memory problems. They also can cause a rare but potentially fatal abnormality in heart rhythms. The SSRIs (sold under the brand names Prozac, Paxil, Zoloft, Luvox, and Celexa) are much safer than the tricyclics and have no dietary restrictions, but they cause sexual dysfunction (loss of sex drive or physical arousal) in many people. Depression itself can cause sexual dysfunction, but SSRIs perpetuate this problem even after the depression is relieved. Newer antidepressants such as Wellbutrin and Remeron do not seem to have the sexual side effects.

Antidepressants take time to work. Even though the drugs boost neurotransmitter levels shortly after being taken, the actual antidepressant effects may take two weeks (more or less) to occur. The boost in neurotransmitters causes the brain to adjust its sensitivity over time, which is probably what really causes the relief from depression. The bottom line is that these drugs cause adjustments in the brain that alleviate depression.

Anxiety and depression occur together much of the time, and they may result from similar problems in the brain. It is not surprising, then, that antidepressant medications also can relieve anxiety. Some studies have shown that some antidepressants are as effective or better than Valium for anxiety. For example, the drug Paxil is now marketed as a treatment for social phobia, which is a form of anxiety.

ST. JOHN'S WORT

St. John's wort *(Hypericum perforatum)* is a perennial herb that produces bright yellow flowers. It was used in ancient Greece and medieval Europe, where it was thought to ward off evil spirits. *Wort* is the Old English word for herb, and it was harvested in Europe on the eve of St. John's day (June 24), when it was burned to purify the air. Traditionally, it was used to treat depression, insomnia, bedwetting, and anxiety.

Active Chemicals in St. John's Wort

St. John's wort contains multiple chemicals that may contribute to its antidepressant effect. The ones that have received the most attention are **hypericin** and **hyperforin.**

Drug Issues with St. John's Wort

St. John's wort should be taken in a concentrated, extract form. But extracts may vary in strength and potency, depending on how they were made and how much of the active chemicals they contain. Standardized extracts should always be used to ensure that the active chemicals are present and in sufficient quantity.

The dosages used in clinical studies of St. John's wort range from 300 to 1,200 mg per day. Studies suggest that St. John's wort should ideally be taken in two or three evenly spaced doses per day. By doing so, steady levels will be maintained in the blood, which gives the best and most consistent effects. A combination of plant chemicals in St. John's wort probably helps it absorb into the body better than if certain chemicals were isolated and taken alone.

How St. John's Wort Works

A good deal of evidence suggests that St. John's wort works in a manner similar to the standard pharmaceutical antidepressants. Namely, it boosts brain levels of the neurotransmitters serotonin, norepinephrine, and dopamine. All known antidepressants are believed to work through a combination of these neurotransmitters. Like the tricyclic and SSRI antidepressants, St. John's wort seems to block these neurotransmitters from being taken back up into neurons, thus prolonging their effects. To a lesser degree, it may prevent the breakdown of these chemicals, also prolonging their effects. Similarly, the continued use of St. John's wort over time causes the long-term brain adaptations that ultimately create the antidepressant effects.

The Effects of St. John's Wort

Sleep

Studies of brain electrical activity during sleep (using the EEG) show that St. John's wort increases the amount of time people spend in sleep Stages 3 and 4. These stages are also called slow-wave sleep, since the brain has minimal activity, and result in a deep, restful sleep.

REM sleep is paradoxically a stage in which the brain is very active, a time when most people experience vivid dreams. For reasons that are not yet known, people with depression go into REM sleep abnormally fast. Similar to other antidepressants, St. John's wort normalizes the amount of time it takes a person to enter REM sleep.

Cognitive Effects

Fortunately, St. John's wort does not seem to have any adverse effects on attention, concentration, or reaction time. This suggests that a person could take it during the day and not have to worry about excessive drowsiness or mental clouding.

Antidepressant Effects

Several studies, in both animals and humans, suggest that St. John's wort has antidepressant effects. The results of multiple human clinical studies have been pooled together to assess the antidepressant effect of St. John's wort. These studies used good scientific methods to test St. John's wort, and had a combined number of 1,757 research subjects. To factor out placebo effects (or expectancy effects), some subjects were given St. John's wort, while others received an identical inactive placebo. Neither the subjects nor the experimenter knew who was getting St. John's wort or placebo until after the study was completed. Consistently, St. John's wort did better than placebo. Thus, people didn't feel better only because they expected to; otherwise both groups would have done equally well on average. Only one recent study failed to support the benefits of St. John's wort in depression. This study (published in the *Journal of the American Medical Association*) had several flaws in the way it was conducted and, suspiciously, was funded and designed by Pfizer, Inc., the makers of Zoloft.

Moreover, hyperforin seems to be an important ingredient in St. John's wort. One study tested two groups: one given an extract high in hyperforin, and another specially prepared to be deficient in hyperforin. The high-hyperforin group showed more improvement in mood. Thus, good-quality brands of St. John's wort should clearly display on the label how much hyperforin is present in the product. Many brands currently label the hypericin content, even though this has been shown to be much less important.

So we know that St. John's wort does something for depression. An important question, then, is how St. John's wort stands up to the pharmaceutical antidepressants. Several studies have investigated this, looking at many different kinds, such as Prozac and Zoloft. In these studies, St. John's wort seemed to do equally well in the right doses. Also similar to the pharmaceutical antidepressants, St. John's wort took about two weeks to really work. Some research suggests that St. John's wort has milder, more tolerable side effects than some pharmaceutical antidepressants as well. Subjects in the controlled studies tend to report fewer and less severe side effects while taking St. John's wort.

Addiction and Dependence

In a world where drug addiction seems to be rampant, it is understandable that people may be highly vigilant, treating all drugs with suspicion. However, absolutely nothing suggests that antidepressant drugs have addicting effects. They do not make a person feel "high" or euphoric the way a drug such as cocaine or heroin might, which is followed by a crash into depression, leading the person to compulsively seek out the drug again. Instead, antidepressants have milder and steadier effects, which are desirable to relieve depression. When people need to stop taking an antidepressant, they are advised by physicians to gradually reduce the dose over a number of weeks, and then stop taking it altogether.

Side Effects

St. John's wort does not seem to have any major side effects in the currently tested doses. In the clinical studies, the number of side effects reported were equal for both St. John's wort and placebo, suggesting that St. John's wort was not the culprit for whatever some research subject happened to be feeling. One study found side effects to be more frequently experienced by those taking pharmaceutical antidepressants (52.8 percent) compared to St. John's wort (19.8 percent). However, we have experience only with short-term use of St. John's wort, so whether it's truly safe for long-term use (e.g., years) is uncertain.

Photosensitivity

Animals who graze on St. John's wort often develop a skin rash in response to prolonged sunlight exposure. A human study of controlled exposure to ultraviolet light found a mild reaction in people taking very high doses (3,600 mg) of St. John's wort. Thus, people taking St. John's wort may want to avoid excessive sun exposure and wear a sun block when outdoors. Since other medications can make people more sensitive to the sun, taking St. John's wort in combination with other medications may worsen this effect.

Interactions with Other Medications

St. John's wort may alter the way other drugs are metabolized in the liver. So far, this is known only for certain other drugs, such as warfarin, digoxin, theophylline, cyclosporin, and indinavir. As with any other medication, people should inform their physicians and pharmacists about all medications they are taking, prescription, over-the-counter, and herbal.

Under no circumstances should St. John's wort be taken along with any other antidepressant medication. This, in effect, is like adding or multiplying the doses and can potentially be fatal. A few cases have been reported, and fortunately the problem was corrected before any serious consequences occurred. People who have bipolar disorder should especially consult a physician before taking St. John's wort.

Sexual Function

Although many SSRI antidepressants seem to worsen sexual function, decreasing the desire and/or ability to perform in both men and women, this does not appear to be the case for St. John's wort. It has not been scientifically studied, but reports so far do not suggest that it has any negative effects on sexual function. This may be a major advantage of St. John's wort over certain other antidepressants.

Conclusions About St. John's Wort

Thus there is good evidence that St. John's wort has a significant antidepressant effect, particularly for mild to moderate depression. Further, it seems to be equivalent to several pharmaceutical antidepressants and may work in a similar way. Like them, it can also take a few weeks before the effects are fully felt. Since it can affect the metabolism of certain other medications, people should discuss the possibility of using it with their doctors, especially if they are already taking medication. It should never be combined with other antidepressants.

GINGER

Ginger comes from the rhizome (or underground stem) of the plant *Zingiber officinale* (see Figure 7.1). It is a tropical plant with long grasslike leaves and cone-shaped flowers at the end. Ginger was well known to the ancient Greeks and Romans, and it was known in China as early as the fourteenth century B.C. The name *Zingiber* derives from the Arabic word *zinschebil,* meaning "root of Zindschi," or India. In ancient herbal traditions, ginger was given to treat a number of ailments, including rhinitis, gingivitis, toothache, painful menstruation, asthma, stroke, constipation, and diabetes. The Spanish used it as early as the sixteenth century, and introduced it to Jamaica and the West Indies, where it is now widely cultivated.

FIGURE 7.1. Ginger *(Zingiber officinale)*. Reprinted from Culbreth DMR (1927). *Materia Medica and Pharmacognosy,* Seventh Edition. Philadelphia: Lea & Febiger.

Preparation

Ginger is most commonly used as a spice. It is harvested eight to nine months after planting. For cooking, ginger is used fresh or dried and ground as a powder. It is also crystallized, candied, and made into syrups. Perhaps one of the most common forms of consumption is the beverage ginger ale, which is made from carbonated water and ginger syrup. "Ginger beer," as it is now commonly called, is merely ginger ale with an increased ginger content. Traditionally, ginger beer was an alcoholic beverage made by fermenting ginger with other spices (e.g., licorice, clove, hops, and gentian). "Ginger Jake" was an alcoholic beverage made during the American Prohibition era. Tragically, it was laced with a toxin that caused irreversible paralysis.

Active Chemicals in Ginger

Several chemicals of interest are found in ginger, which may vary according to the plant's maturity when harvested, and the climate and soil in which it's grown. The primary pungent chemical in ginger is called **gingerol.** Other closely related chemicals found are **shogaol** (from the Japanese name for ginger, *shoga*) and zingerone, which only are present in small amounts in fresh ginger, but increase when it is stored. Another group of chemicals in ginger, called terpenoids, include zingiberene, zingiberol, and galanolactone.

How Ginger Works

Ginger's effects on anxiety are uncertain as the few possibilities are not very well understood at present. Ginger does seem to work at receptors for the neurotransmitter serotonin, which could give it anti-anxiety effects, but this remains to be proven.

Effects of Ginger

Blood Clotting

Ginger is known to reduce blood clotting in a manner similar to drugs such as aspirin. This has been demonstrated in humans who took oral doses of ginger. But, while fresh ginger had this effect, dried ginger did not. Danger to people taking blood thinners (anticoagulant medications such as aspirin and warfarin) isn't likely, particularly since ginger is usually consumed dried or cooked.

Digestive Effects

Ginger has long been reputed to reduce nausea. Research is still being conducted, but so far several studies support its effectiveness. This includes nausea induced by several causes, including general anesthesia for surgery, seasickness, morning sickness, and chemotherapy-induced nausea. In people with seasickness, ginger also reduced cold sweats.

Anti-Anxiety Effects

Two studies have investigated anti-anxiety effects of ginger. However, they used only animals, so it remains to be proven effective in humans. Also, the ginger was given only in combination with ginkgo, so it's impossible to know whether ginger or ginkgo was the necessary ingredient, or perhaps both are necessary. The ginger/ginkgo combination was effective in reducing anxiety in rats, but they had to be given in the right proportions (2.5 parts ginger to 1 part ginkgo). Unlike Valium, which was used for comparison, the ginger/ginkgo combination did not impair memory.

A study of shogaol in rats showed that it slowed brain electrical activity on the EEG. This would be consistent with a drug that might reduce anxiety, but does not prove it's effective. Human studies are still needed to show whether ginger, or a ginger/ginkgo combination, will reduce anxiety.

Side Effects of Ginger

Ginger seems to have few side effects when taken in normal amounts. The Food and Drug Administration includes ginger on the GRAS (Generally Regarded As Safe) list. Ginger can cause a warm or burning sensation in the stomach, which can be relieved by drinking a mild beverage. The paralysis caused historically by the alcoholic beverage Ginger Jake was due to a contaminant called TOCP, and had nothing to do with the ginger itself.

GINKGO

Ginkgo is generally known for its effects on memory and thinking, but it may have psychotherapeutic uses as well. It is discussed more extensively in Chapter 5, but mentioned here for possible antidepressant and anti-anxiety effects.

Anti-Anxiety Effects

As mentioned, a ginger/ginkgo combination was shown to reduce anxiety in rats, but human studies have not yet been done and it is not clear whether either of the two drugs would be effective by itself. Other animal studies of ginkgo alone showed that it reduced memory problems caused by stress. These studies are promising, but need to be repeated with human subjects.

Antidepressant Effects

One study examined ginkgo as a treatment for seasonal affective disorder, or winter depression. Ginkgo was not effective for this use, although it's possible that a higher dose or starting the treatment earlier (before the depression started) may have changed the outcome.

Although it may not be an antidepressant by itself, one Chinese study suggests that adding ginkgo to a pharmaceutical antidepressant may improve its benefits. One group of subjects took the standard antidepressant amitriptyline plus placebo, and the second group took amitriptyline plus ginkgo. The amitriptyline plus ginkgo group did better. People with depression typically experience poor-quality sleep. Even though they may get a full night's sleep, they are likely to feel poorly rested and may wake up very early in the morning, unable to return to sleep. A recent study used the EEG to measure the effects of ginkgo on the sleep of depressed people. It showed that ginkgo improves deep, restful sleep and reduced the number of awakenings.

Thus, ginkgo may not be the sole choice for treatment of depression, but it seems to be a useful addition to someone receiving other forms of treatment.

How Ginkgo May Help Depression

The most likely mechanism by which ginkgo improves depression is through stress hormones (see Figure 7.2). When someone experiences emotional stress, a part of the brain called the hypothalamus releases a chemical that signals the pituitary gland. The pituitary gland, in turn, releases another chemical that travels through the blood to the adrenal glands, which are situated on top of the kidneys. The adrenal glands release hormones such as cortisol, which help prepare the body and brain for the stressor. In a nondepressed person, this hormonal cascade occurs when stress begins and then subsides when the stressor is gone. In a depressed individual, however, the system can't seem to turn itself off. The stress hormones are persistently released. This probably perpetuates the depression and eventually has toxic effects in the brain. Ginkgo has been shown to reduce the production of stress hormones. Thus it may normalize levels of stress hormones in people with depression, which could affect mood.

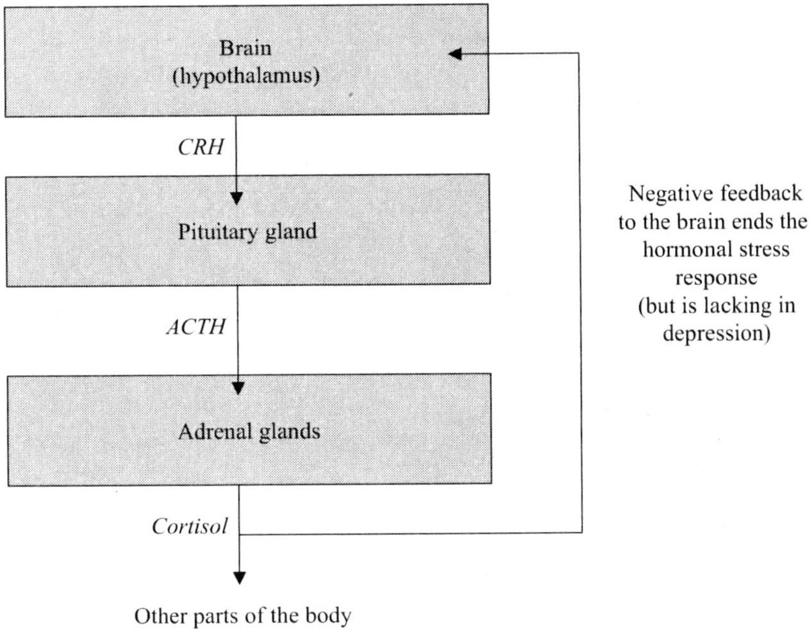

FIGURE 7.2. Stress hormones in the brain. A chain of hormone releases occurs with stress, starting with the hypothalamus, then the pituitary and the adrenal glands. Normally, the hormones feed back to the brain, eventually shutting down the stress hormone release. In depression, the brain becomes less sensitive to the negative feedback effect, failing to shut down the stress hormone release.

SCHIZOPHRENIA AND PSYCHOSIS

Schizophrenia is often used to refer to a group of psychotic disorders in which the person suffering has difficulty distinguishing reality. The symptoms of psychotic disorders are thought of as positive and negative symptoms. Positive symptoms are excesses, or things that are present and should not be, such as hallucinations and delusions. Hallucinations are defined as perceiving something that isn't really there. In schizophrenia, the most common hallucinations are auditory, such as hearing voices. Delusions are false beliefs, such as paranoid thoughts that people are plotting against you when that is clearly not the case. Negative symptoms, on the other hand, are things that are lacking. For example, people with schizophrenia have cognitive deficits such as impairments in logical and abstract thinking, paying attention, organizing thoughts, expressing emotions, and social situations.

Pharmaceutical Antipsychotic Medications

Many standard **antipsychotic medications** are available. They all seem to work by one common mechanism: blocking receptors for the neurotransmitter dopamine. This makes the brain less sensitive to the neurotransmitter and reduces the positive symptoms of schizophrenia. Evidence suggests that people with schizophrenia are overly sensitive to dopamine. Making even a normal, nonschizophrenic person oversensitive to dopamine can cause hallucinations, which is what happens when people take large doses of cocaine or amphetamine.

Blocking dopamine reduces the brain's sensitivity and reduces hallucinations and delusions in people with schizophrenia. In contrast, traditional antipsychotic medications (such as Haldol and Thorazine) do not seem to help the negative symptoms at all. A newer generation of antipsychotic drugs (called **atypical antipsychotics**) are available that block not only dopamine, but also certain serotonin receptors. These include the drugs Risperdal, Zyprexa, and Seroquel. These have the additional benefit of reducing some negative symptoms as well.

Side Effects of Pharmaceutical Antipsychotic Drugs

Antipsychotic drugs have many undesirable side effects, which unfortunately make them unappealing for those who need them. Older medications produced abnormal, involuntary movements, but newer medications have reduced this problem. In some cases, long-term use can cause permanent involuntary movements called **tardive dyskinesia.** However, this only occurs in about 15 to 25 percent of chronic psychotic patients.

Herbal Medicines for Schizophrenia

Herbal medications are not a good choice as the sole treatment for schizophrenia. Serpentwood is described here for historical reasons, but modern antipsychotic drugs are a better choice. However, ginkgo added to antipsychotic medications may help reduce some of the side effects of the medication.

Serpentwood

Serpentwood *(Rauwolfia serpentina)* grows as a flowering small tree or shrub. It has had a long traditional reputation as a treatment for schizophrenia in India, where it is known as *chandrika*. In the West, medical interest in the plant began in the 1950s, when the chemical **reserpine** was shown to reduce blood pressure. The medicinal part of the plant is the dried root.

How serpentwood works. The main psychoactive drug in serpentwood is a chemical called reserpine. Reserpine basically prevents the storage of the neurotransmitters serotonin, dopamine, and norepinephrine in the brain. When they aren't stored, the neurotransmitters fall prey to breakdown by enzymes, depleting the brain of those chemicals.

Serpentwood reduces the amount of available dopamine in the brain. Since people with schizophrenia are oversensitive to dopamine, this partially corrects the problem. So the historical reputation for serpentwood is probably well deserved. Nonetheless, modern pharmaceutical antipsychotic drugs are better choices and produce fewer side effects.

In the autonomic nervous system (which controls automatic functions such as heart rate, digestion, and salivation), serpentwood has the effect of lowering blood pressure. Reserpine, and similar drugs developed later, were enormously successful in treating high blood pressure. Since we started using them, there has been a tremendous reduction in the number of strokes in the United States. However, this particular type of medication has undesirable side effects, and better blood pressure medications have since been developed.

Side effects of serpentwood. As discussed in the antidepressant section, the neurotransmitters serotonin, dopamine, and norepinephrine are important for regulating mood. Antidepressant drugs boost these chemicals, and depleting the brain of them can cause depression. This is the primary problem with serpentwood and its active drug, reserpine. People who have experienced depression are at greatest risk, and the problem may often go unrecognized. The same neurotransmitters are also important for many thinking abilities, such as concentration and short-term memory. Serpentwood has negative effects on these abilities as well. Given these important side effects, serpentwood is best not used for schizophrenia or high blood pressure.

Ginkgo

Two recent clinical studies have shown that ginkgo may be of use to people with schizophrenia. These studies show that schizophrenic people who took the antipsychotic drug Haldol plus ginkgo did better than those who took Haldol plus placebo. A greater percentage of those taking the additional ginkgo showed a beneficial response to the Haldol (57.1 percent), as compared to those who took the Haldol alone (37.7 percent). The improvement seemed to be mostly in reduction of positive symptoms. These findings are preliminary and need further research, but this could be yet another clinical use of ginkgo.

Chapter 8

Herbal Painkillers

Pain could be described as a necessary evil. *Evil* describes it because it is the source of so much human suffering, to the extent that the word *pain* is often used synonymously with suffering. *Necessary* also describes pain because it serves such an important function in our lives—without it, we would surely maim and kill ourselves before long (which actually happens in some unfortunate individuals who lack pain). Although necessary, it's only useful up to a point. When pain outlives its usefulness, it hinders us. Accordingly, plants that have the ability to alleviate pain are very important to us.

WHAT IS PAIN?

We can examine pain on several different levels (see Table 8.1). At the most basic level, pain is a sensation. It is information from the body that is carried through nerves to the brain, telling us that some sort of damage is occurring. But pain is also an emotional phenomenon. It inflicts suffering. When we experience pain, we don't just passively notice it, but instead get a strong feeling such as distress, anger,

TABLE 8.1. Aspects of pain.

Aspect	Example
Sensory	Bodily feeling of pain
Arousal	Alertness in response to pain
Emotional	Suffering, distress in response to pain
Cognitive	Situational context of pain Cause of pain Implications of pain

or maybe even exhilaration. That strong emotion motivates us to do something about it, usually to get away from whatever is causing the pain. Pain also causes alertness and arousal, to focus our attention on its source. A perfect example of this would be stubbing your toe while walking to the bathroom in the middle of the night—from a half-asleep state, you're suddenly wide awake. Pain also has a cognitive aspect, a meaning and context that colors the experience. A person dying from cancer and a woman giving birth to a child may experience the same intensity of pain, but the pain has different meanings. For one it means the beginning of a new life, whereas for the other it threatens an untimely end.

Pain Relief

Pain, by nature, is a subjective experience, but each person experiences it privately in the mind. The only truly objective aspects of it are measurements of nerve cell activity corresponding to pain, or behaviors in response to pain. We may see someone bump his arm in a doorway, but we wait to see his reaction to know whether it actually hurt.

The relief of pain is called **analgesia.** It is created by painkiller drugs such as aspirin and morphine. They selectively block pain, but not other bodily sensations such as temperature or touch. **Anesthesia,** on the other hand, means blocking all sensations, pain and otherwise. Although anesthesia blocks all sensation, its main purpose is really to stop pain. There are two types of anesthesia: local anesthesia is caused by drugs that stop nerve signals from reaching the brain (as when a dentist numbs your mouth before drilling a tooth), while general anesthesia deeply sedates a person to maintain unconsciousness during surgery.

THE PAIN SYSTEMS

To appreciate painkillers, or analgesic drugs, it's helpful to understand a little about how pain is transmitted to begin with. Let's take an example of pain and follow its course (see Figure 8.1). Suppose you're cooking a meal and accidentally touch a hot pot. The intense heat damages your skin, burning a spot. A set of nerves in the skin is

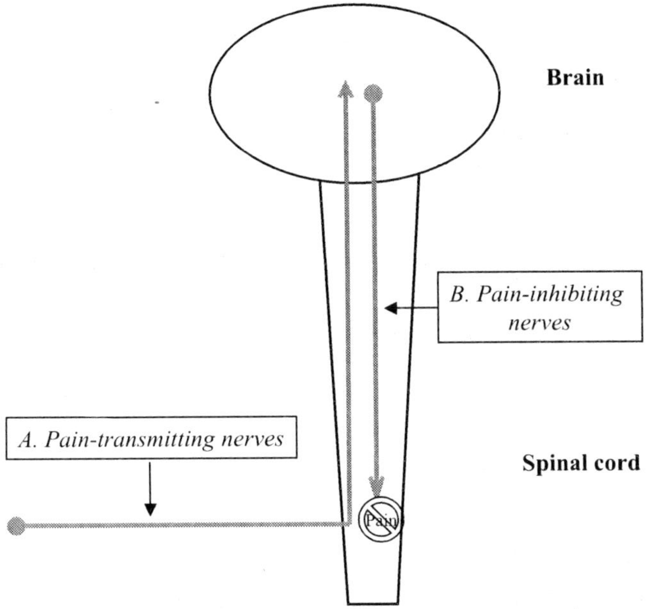

FIGURE 8.1. Pain and analgesia systems. Painkillers can either reduce pain signals reaching the brain (A), or dampen pain once it has already reached the brain (B).

specialized to transmit pain signals. In the burned area, the pain-transmitting nerves are activated. Normally they sit dormant, since we do not experience pain most of the time. But whenever any type of intense stimulation occurs (whether from heat, pressure, or even a chemical), the pain nerves activate. Chemicals are released in the damaged region that make the pain nerves even more sensitive, ensuring that the important message will be sent. An electrical signal passes along the nerves, reaching the spinal cord, and some pain-signaling neurotransmitters are released. This activates a set of pain nerves that travel up the spinal cord to the brain, where the message is relayed to several areas. For example, the pain nerves going to your parietal cortex convey the sensation of pain, letting you know where on your body the burn occurred. Other signals travel to the limbic system, causing an emotional reaction such as surprise and distress. Pain signals reaching the reticular formation are responsible for the seemingly instantaneous alertness we experience with pain.

A number of chemicals and neurotransmitters are involved in the pain transmission process—too many to mention all of them here. But to give a few examples, a chemical called **prostaglandin E** is one of the messengers released in the skin when damage occurs, making the pain nerves more sensitive. Substance P is a neurotransmitter released in the spinal cord that passes the signal to neurons going up the spinal cord to the brain. Each of these chemicals represents an opportunity to block the pain with a drug.

THE ANALGESIA SYSTEM

Most people intuitively know that they have a pain-transmission system, but few people know that the nervous system contains a pain-inhibition system too. Even though pain is useful and necessary, too much pain is incapacitating. So the brain has its own system for dampening pain.

Just as the pain nerves travel up to the brain, the **analgesia system** passes them on the way down. It starts in the brainstem, mainly, and travels down to the spinal cord. When our pain nerves are active, the analgesia system is working in the spinal cord to tune it down. It may not seem that way, because we still experience pain, but if we didn't have this system, injuries would hurt much more. We've all heard stories of people experiencing serious injuries, such as in an accident, and not even realizing they were hurt until afterward. That's because the analgesia system is temporarily stopping the pain.

Many neurotransmitters are involved in the analgesia system, including serotonin, norepinephrine, acetylcholine, and the opioids, to name a few. The opioids, such as endorphins, enkephalin, and endomorphin, are a class of neurotransmitters that have attracted a lot of attention for their role in stopping pain. Again, each one of these neurotransmitters represents another potential opportunity for a drug to stop pain.

Some Disorders of Pain

At best, pain warns us of danger, and when the danger is gone, the pain subsides. If only it were always so simple. In many conditions, this process goes astray (see Table 8.2). Chronic pain is probably the most common pain disorder. It seems the mind and brain are not well equipped to handle incessant and severe pain. Like an alarm that is constantly going off, chronic pain robs people of their resources, both cognitive and emotional. By excessively drawing on mental resources, pain causes abilities such as attention and memory to suffer. By overstimulating emotional areas of the brain, pain commonly causes depression. Psychotherapy is very useful in helping people with chronic pain find strategies for coping and getting on with their lives.

Hyperalgesia is a pain disorder involving an abnormal oversensitivity to pain. People so afflicted may put on a sweater, but feel as if steel wool is dragging over their skin. On the other end of the spectrum is a rare condition called **congenital insensitivity to pain** (sometimes called congenital analgia). Without the deterrent warning signal of pain, these children bite off the tips of their fingers and tongues, carelessly cut and burn themselves, and often break their bones. People with another disorder, called **pain asymbolia,** are able to sense pain, but have no emotional reaction to it. They feel the pain, but it does not cause them any distress. It is easy to see that either too much or too little pain is harmful.

HERBAL ANALGESICS

Many plants have analgesic and anesthetic effects. Even though they are different plants, some of them block pain through similar chemical effects. Here, the analgesic herbs are grouped with those that have similar actions and effects (see Table 8.3). They all work in two very general ways: they can either stop a pain message from get-

TABLE 8.2. Disorders of pain.

Disorder	**Symptom**
Chronic pain	Persistent or recurring pain, incapacitating
Hyperalgesia	Oversensitivity to pain
Congenital analgia	Lacking the sensation of pain since birth
Pain asymbolia	Feeling the sensation of pain, but lacking an emotional response

TABLE 8.3. Summary of analgesic herbs.

Herb	Active chemical	How it works
Opium-like analgesics		
Opium poppy	Morphine, codeine	Activates analgesia system
Myrrh	Furanosesquiterpenes	Activates analgesia system
Aspirin-like analgesics		
Willow	Salicylate	Makes pain nerves less sensitive
Meadowsweet	Salicylate	Makes pain nerves less sensitive
Feverfew	Parthenolide	Makes pain nerves less sensitive
Ginger	Shogaol, gingerol	Makes pain nerves less sensitive
Burning analgesics		
Chili	Capsaicin	Makes pain nerves less sensitive (over time)
Tobacco-like analgesics		
Tobacco	Nicotine	Activates analgesia system
Lobelia	Lobeline	Activates analgesia system
Areca	Arecoline	Activates analgesia system
Caffeine-containing analgesics		
Coffee	Caffeine	Blocks pain signals
Tea	Caffeine	Blocks pain signals
Cocoa	Caffeine	Blocks pain signals
Maté	Caffeine	Blocks pain signals
Cola	Caffeine	Blocks pain signals
Guarana	Caffeine	Blocks pain signals
Amphetamine-like analgesics		
Ephedra	Ephedrine	Activates analgesia system
Khat	Cathinone	Activates analgesia system
Coca	Cocaine	Activates analgesia system
Cannabis analgesia		
Cannabis	THC	Activates analgesia system

ting to the brain, or they can activate the brain's analgesia system to dampen pain signals that have already arrived.

OPIUM POPPY

The opium poppy *(Papaver somniferum)* is a blue-gray plant that grows about 1 to 5 feet in height (see Figure 8.2). At the top of its stem is a round seed pod that resembles a miniature pumpkin. The seeds are small, kidney-shaped—the familiar black dots on top of a poppy roll or in a poppy pastry. Opium is the juice collected from the seed pod, which contains the active drugs.

The opium poppy is one of a few plants that has helped shape human history. As a painkiller, it was unmatched for millennia. As a drug that makes people feel soothed and placid, it has been a source of addiction across uncounted generations. Poppy seems to have been used by humans for most of history. Archaeologists have found poppy seeds in villages dating back to the Neolithic period and Bronze Age. The ancient Greeks, Romans, and Egyptians were very familiar with poppy's effects. Ancient Arabs introduced poppy to Asia, where it was used to treat dysentery. Smoking opium, which has greater addictive potential, became widespread in Asia during the

FIGURE 8.2. Opium poppy *(Papaver somniferum)*. Copyright 2004 by Marcello Spinella.

1700s. In the United States, opium products were common in over-the-counter medicines at the beginning of the twentieth century. Bayer (the aspirin company) created **heroin** from morphine, and sold it as a cough syrup. As addiction became a widespread problem, laws were enacted to restrict its use.

Opium poppies are grown in a large tract of land extending from Turkey to Southeast Asia. Because growing opium is so lucrative, this area is called the Golden Crescent. Opium is extracted by making razor cuts on the seed capsules at the right time of development. The cuts exude a milky white sap, which turns red-brown with exposure to the air. This raw opium is collected, dried, and compressed.

Active Chemicals in Opium

More than thirty active chemicals (called alkaloids) have been identified in opium. The two with analgesic effects are morphine and codeine. Morphine was discovered in 1810, when it was named after Morpheus, the Greek god of dreams. The pharmaceutical morphine and codeine we use are still refined from opium because it's cheaper and easier than making them synthetically. Heroin, or diacetylmorphine, is chemically modified morphine that is much more potent, ounce for ounce.

There are many **opioid** (meaning opium-like) drugs, which have effects similar to morphine and codeine. These include meperidine (Demerol), propoxyphene (Darvocet), and methadone. These are examples of synthetic drugs created in the twentieth century, and do not derive from any plant. However, they were specifically designed to imitate the effects of morphine and codeine.

How Opium Works

Morphine and codeine work by imitating the opioid neurotransmitters. The opioid neurotransmitters are naturally produced in the nervous system, and play many roles, including suppression of pain. Morphine and codeine imitate these neurotransmitters by stimulating their receptors. This has the effect of temporarily dampening or shutting off the transmission of pain signals, before they travel up the spinal cord.

Drug Issues with Opium

Opioid drugs are taken in many ways. As with many drugs, more purified forms and more intense ways of administering them have become available over time. During the eighteenth century, raw opium was commonly smoked. The invention of the hypodermic needle gave people a more powerful way to administer purified morphine, for medical use or otherwise. By the mid-1800s, injection of morphine for severe pain was well known to medical doctors, and it was widely used during the American Civil War. Although statistics were not kept at the time, addiction to morphine among war veterans was considered a serious problem. The creation of heroin from morphine increased its potency, allowing even more intense psychological effects than morphine could cause. Heroin dissolves better in fat than morphine does, which makes it more able to gain access to the brain. So, with equal amounts of morphine and heroin, more heroin penetrates into the brain and has a greater effect. Since smoking a purified drug is the most direct way of delivering it to the brain, smoking heroin (known as "chasing the dragon") has become one of the most dangerous drug practices.

Since the commonly eaten poppy seeds come from the same poppy plant, concern arose that the seeds may erroneously cause a positive result on drug tests. The amount of morphine and codeine in poppy seeds is very small, but consuming enough of them (as in a poppy strudel) could cause a false-positive drug test. However, there is a way to tell heroin and poppy seeds apart chemically. Heroin is converted into a chemical called 6-MAM, which is not found in poppy seeds. Poppy seeds contain a chemical called thebaine, which is not found in heroin. Thus, someone who tests positive for 6-MAM must have gotten it from heroin, whereas only someone who ate poppy seeds would test positive for thebaine.

Effects of Morphine

Analgesia

Morphine and codeine are potent analgesics. They block the sensation of pain, and they also dull the emotional aspect of pain. So a person not only feels less pain, but also cares less about whatever pain may still be felt. Continuous, dull pain is the most responsive to opioid drugs, while quick, sharp pain is less affected by opioids. Pain from tissue damage (e.g., surgery) responds well to opioids, but pain from a nerve injury (neuropathic pain) does not.

Emotional Effects

Morphine and codeine create a pleasant, euphoric state of mind. Anxiety is reduced or abolished, creating feelings of contentment and complacency. These responses are largely due to their effects in emotional regions of the brain, the limbic system. The intensely pleasurable effects are a chief reason for the addictive potential of opioid drugs. Many people have found their effects too appealing.

However, opioids have also been unnecessarily demonized. When people use these drugs under medical supervision, for a limited time, and to treat an illness, the vast majority of people do not become addicted. The drug serves its purpose and then may be discontinued. Opioids are powerful and useful medicines, and when used appropriately they alleviate suffering.

Cognitive Effects

Opioids cause sedation, but not as strongly as sedative drugs such as barbiturates or general anesthetics. Although they make a person sleepy, the person can be woken up (except for overdoses). They slow the brain's ability to process information, and they impair memory. They do not cause complete memory loss, but people on opioids are not at their mental peak.

Cardiovascular Effects

Morphine can cause orthostatic hypotension, which is a drop in blood pressure when a person stands suddenly that can cause lightheadedness. To avoid this, standing up quickly should be avoided when taking these drugs.

Effects on Breathing

Most opioid drugs suppress breathing. Normally the brain is sensitive to carbon dioxide levels in the blood. If they get too high, we start breathing more to correct them. Opioids make the brain less sensitive to carbon dioxide, so breathing is reduced. In the most extreme cases, such as an overdose, this can stop breathing altogether.

Digestive System Effects

Opioid drugs generally slow the digestive process. Muscle contractions that move food through the digestive tract are stopped. As a result, constipation is a common problem with opioid use. Morphine and heroin initially can cause nausea and vomiting, but this is due to stimulation of a brainstem area that triggers vomiting (called the area postrema).

Cough

Opioids are very effective at suppressing coughs. This is also due to suppression of a brainstem area that triggers coughs. Cough suppressant was a major use of opioids, but their addiction potential limited widespread use.

Addiction and Dependence

The intense emotional effects of opioid drugs give them abuse potential. A core feature of drug addiction is that people use the drug maladaptively to manage mood, alleviating depression and calming fears and anxiety. Opioids have both of these effects, lifting mood while simultaneously making people feel placidly calm. The chemistry of addiction is complex, but like many drugs of abuse, opioids have an impact on the neurotransmitter dopamine. As stated, these addictive effects are also evidenced by opioids' stimulation of emotional areas of the brain. Brain activity scans of people with opioid addiction show that their brains function somewhat differently than others'. For example, a lack of brain activity in some areas (e.g., in the temporal lobe) was associated with depression in this population.

Opioid drugs are infamous for their ability to cause dependence. If a person uses opioids chronically and suddenly stops, a withdrawal syndrome occurs. The intensity of the withdrawal syndrome varies with how much was used and for how long, but it can become intense and very unpleasant (see Table 8.4). Understandably, it is a deterrent to many people who would like to quit using heroin but cannot endure the process.

Drug Treatment of Opioid Addiction

Many drug strategies treat opioid addiction. The most common is a replacement strategy. The drug **methadone,** which is less intense and longer acting than heroin, keeps users in a more steady state, rather than the repetitive fluctuations of high and withdrawal that happen with heroin. Methadone can then be gradually reduced, so that only the milder withdrawal symptoms are experienced (but for a longer period of time). Alternately, methadone maintenance can be done, which keeps the person on methadone indefinitely. A newer drug called **LAAM** (L-alpha-acetylmethadol) is being investigated as an alternative to methadone.

The drug clonidine (a blood pressure medication) is also used to treat opioid addiction. Although it works in a completely different way than the opioids, it does reduce the severity of symptoms, making the process a little more tolerable. Another drug, called **naltrexone** (Trexan), can be used after the withdrawal period is done. This drug blocks opioid receptors, nullifying the effect of any morphine or heroin taken thereafter. So, a daily Trexan tablet can help people who want to remove the temptation of relapsing, since any opioid drug used afterward would be ineffective.

The strategies for opioid treatment really only address the short-term aspect of addiction. Although withdrawal is done within seventy-two hours, people may feel depressed and crave the drug for

TABLE 8.4. Opioid withdrawal syndrome.

Time lapse	Symptoms
8 hours	Tearing, yawning, sweating, and runny nose
12-14 hours	Tossing, restless sleep
48-72 hours	Loss of appetite, nausea, vomiting, diarrhea; goose-flesh, rapid heartbeat, tremors, muscle cramps, chills and flushing, restlessness, irritability, severe anxiety, depression

months or even years afterward. If a person does not get any treatment after withdrawal, the chances of relapse are high. With any good addiction treatment, counseling is absolutely necessary. The cycle of addiction can only be broken when people are aware of their behavior patterns and underlying reasons for drug use, and have acquired strategies to avoid these problems in the future.

Side Effects of Opioids

The most dangerous aspect of opioid drugs, particularly when they are injected intravenously, is the chance of overdose. When given an overdose, the person often becomes unconscious and cannot be fully wakened. Breathing slows and may stop entirely, which is the main cause of death with overdoses. The skin becomes cold and clammy and death can result quickly from a lack of oxygen. The invention of drugs such as naltrexone (Trexan) allows emergency medical workers to rapidly reverse this condition if the person is reached in time. A person who is slipping into a coma one minute can literally be awake the next. But as the brain is very sensitive to oxygen deprivation, the window of opportunity is narrow. The other danger of this condition applies to people who use opioids regularly: when the Trexan reverses the overdose, it can also instantly throw them into a full-blown withdrawal syndrome. When this happens, people must be watched closely to ensure that their vital functions are stable.

Prenatal Effects

Long-term use of opioid drugs during pregnancy causes dependence in the fetus, who undergoes a withdrawal syndrome after birth. Opioid abuse is associated with lower birth weight and increased risk of mortality in newborns.

MYRRH

Myrrh *(Commiphora molmol)* is a shrub that grows in eastern Mediterranean and Middle East regions. A resin exuded from the bark historically was made into an ointment. Its use was known to the ancient Egyptians, Jews, Romans, and Greeks. In the Bible, *vinum* *murratum,* a wine containing myrrh, was reputedly offered to Christ before the Crucifixion.

Active Chemicals and Effects of Myrrh

Myrrh contains chemicals called **furanosesquiterpenes.** In studies, when these chemicals were given to mice, they had analgesic effects. The analgesic effects were prevented by the drug naloxone, which means they seem to involve the opioid system. So, myrrh may have painkilling effects through similar systems. It doesn't seem likely that myrrh will be used as a painkiller clinically, but it is interesting to understand why the plant was used historically.

WILLOW

Several trees belong to the genus *Salix* (see Figure 8.3). The American **willow** *(Salix nigra)* is a tree that grows 20 to 60 feet in height. It has gray bark and small, spearhead-shaped leaves. The part of the tree

FIGURE 8.3. White willow leaves *(Salix alba).* Copyright 2004 by Marcello Spinella.

used medicinally is the bark. Willow bark has been used throughout recorded history to treat pain, fevers, and inflammation. Its uses were discovered independently in many diverse cultures such as those of the ancient Greeks, Egyptians, Hottentots, and American Indians.

Active Chemicals in Willow

The active chemical in willow was isolated in 1829 and named salicin after the Latin name for willow, *Salix*. When taken, salicin is converted in the liver and intestines into the actual active drug, salicylic acid (or **salicylate**). Salicylic acid causes stomach irritation, but modified as acetylsalicylic acid (often abbreviated ASA), it is better tolerated. Acetylsalicylic acid is better known by its common name, aspirin.

Aspirin was the most commonly used analgesic drug until the invention of the similar drugs acetaminophen, ibuprofen, and others. Collectively, these drugs are known as the NSAIDs (nonsteroidal anti-inflammatory drugs). Between 10,000 and 20,000 tons of aspirin are consumed annually in the United States.

Drug Issues with Willow

Since salicin occurs in small amounts in willow, raw willow is impractical to use. Consumed as a tea, it would take between three and twenty cups to get an effect, and at such high doses, chemicals called tannins in the bark would make the user sick. Thus, a willow extract or aspirin is needed for therapeutic benefits.

A study of willow bark extract in humans showed that 240 mg of salicin in an extract would be equivalent to about 87 mg of aspirin. Actual aspirin is absorbed very well into the body, reaching significant blood levels in thirty minutes. Aspirin absorbs well through the skin as well, allowing it to be used in a topical cream.

Salicylate, aspirin, and other NSAIDs work by inhibiting the production of chemicals in the body called prostaglandins. Prostaglandins are created with the enzyme **cyclooxygenase,** and they make pain nerves more sensitive. By blocking cyclooxygenase, salicylate stops pain nerves from becoming too sensitive.

The Effects of Willow

Salicylate and related drugs are known to have four main effects: reducing pain (analgesic), fever (antipyretic), inflammation (anti-inflammatory) and blood clotting (anticoagulant).

NSAIDs are effective for low- to medium-intensity pain but not for high-intensity pain. They work for headache and muscle pain, but are less effective for toothaches and sore throat, and ineffective for deep visceral and traumatic pain. Although the analgesic effects of aspirin are well known, much less research has been done on willow extract. However, a few studies suggest that it is effective, especially in doses standardized for 240 mg of salicin. Clinical studies of willow bark extract support its use in osteoarthritis and lower back pain.

NSAIDs lower fevers produced by infection. Normal body temperature is not lowered by aspirin and it has no effect on heat created by exercise. When the body contracts an infection, chemicals called cytokines are released in the bloodstream, which enter the brain and cause the production of prostaglandins in the hypothalamus. This in turn causes the fever, which can be reversed by NSAIDs.

Aspirin is recognized as an important blood-clot inhibitor. Willow extract has the same effect, but to a somewhat lesser degree. Aspirin has been called the most useful drug in the prevention of stroke, which can result from blood clots in the brain. Several controlled clinical studies show that aspirin significantly reduces the risk of stroke. Willow extract has not been tested in this regard.

Side Effects of Willow

Overall, aspirin is fairly low in toxicity. It produces little tolerance and no addiction, so it can be used over time. However, in high doses (10 to 30 g) it can be fatal. High doses cause headache, dizziness, ringing in the ears (tinnitus), dimmed vision, hearing difficulties, confusion, sweating, nausea, hyperventilation, and diarrhea. Even higher doses cause metabolic disturbances, bleeding, seizures, and coma.

Reye's syndrome is a rare but serious condition that can result from aspirin use. It has been noticed primarily in people under age twenty who took aspirin during a viral infection, especially influenza or chicken pox. Victims become disoriented, their behavior becomes unusual, and they may vomit continuously. Permanent brain damage may occur, and approximately 30 percent of victims have died. Fortunately, since it was first reported in 1980, the number of cases has decreased steeply. An evaluation of many reported cases found they were actually other disorders misdiagnosed as Reye's syndrome.

MEADOWSWEET

Meadowsweet *(Filipendula ulmaria)* is another plant, like the willow species, which produces salicylates. Thus, the way it works and its effects are similar to those discussed for willow. Willow has a greater amount of salicylates, so meadowsweet has milder effects and is less practical to use. Although it has been used historically and we understand the plant's chemistry, there have not been any controlled studies of meadowsweet in humans.

Incidentally, meadowsweet is the plant from which aspirin got its name. Meadowsweet was known as *Spiraea ulmaria* at the time aspirin was invented. The *a* comes from acetyl, and *spir* from *Spiraea*. The Bayer company started selling aspirin in 1899, around the same time it started selling its other painkiller, heroin.

FEVERFEW

Feverfew *(Tanacetum parthenium)* has a reputation for treating pain, migraine, arthritis, fevers, and anxiety. It is an aromatic flowering herb, which is daisylike in appearance. Feverfew originated in southeast Europe, but now grows in Australia and North America. It can be taken fresh by chewing the leaves, but is also available as a standardized extract.

Active Chemical in Feverfew

The main active chemical constituent in feverfew is **parthenolide,** a sesquiterpenoid lactone. Like aspirin, it inhibits the enzyme cyclooxygenase and the production of prostaglandin. Parthenolide also has a very different effect, stopping the effects of serotonin by blocking serotonin receptors and maybe preventing its release.

Effects of Feverfew

An animal study showed feverfew to have analgesic and anti-inflammatory effects. However, these effects have not been carefully studied in humans. Only one study was done in people with rheumatoid arthritis. It was not effective in this study when given over a six-week period. However, this study used chopped leaves and did not control the exact dose of parthenolide given to subjects. The fact that parthenolide blocks cyclooxygenase and reduces prostaglandins suggests that it has some potential. The analgesic and anti-inflammatory effects of feverfew need to be more carefully studied in humans.

However, there is some good evidence for antimigraine use of feverfew in humans. It appears to reduce the number and severity of attacks, as well as the vomiting caused by the attacks. The most likely candidate for this effect is feverfew's impact on serotonin receptors. Migraines result from abnormal blood flow to the brain and its coverings (the meninges). Serotonin regulates blood flow, and many pharmaceutical antimigraine medications work through serotonin receptors. Although feverfew seems promising in this regard, more research needs to be done.

As its name suggests, feverfew may reduce fevers caused by infection. Again, its effects on cyclooxygenase and prostaglandins suggest that feverfew could be effective in this respect. However, studies have not yet been done.

Although traditional use of feverfew suggests possible anti-anxiety effects, this has not been tested in either animals or humans. Some drugs that block serotonin receptors reduce anxiety, so it may be true. This is yet another untested possibility of feverfew.

Side Effects of Feverfew

No major adverse effects are known to result from typical doses of feverfew. Since feverfew inhibits blood clotting, it should not be taken by people already taking another blood thinner such as aspirin or warfarin. When taking feverfew over time, it should be slowly and gradually discontinued. Abruptly stopping it can cause rebound symptoms, such as migraines, anxiety, poor sleep, and stiffness of the muscles and joints.

GINGER

Ginger comes from the plant *Zingiber officinale*. It was discussed in more detail in Chapter 7, but is mentioned here for its NSAID and analgesic effects.

How Ginger Works

Like other NSAIDs, chemicals in ginger prevent the formation of certain prostaglandins. A recent study has shown chemicals from ginger to be strong inhibitors of cyclooxygenase. Some of the chemicals that have these effects include gingerdione, shogaol, and gingerol. The word shogaol comes from the Japanese word for ginger, *shoga*.

Effects of Ginger

Ginger was shown to have analgesic effects in animals in a study published in 1984. Controlled human studies were slow to follow, but a few have been done. Two recent studies suggest an analgesic effect of ginger extract. A controlled study in 2000 on people with osteoarthritis first hinted at the possibility. A more recent study, also in people with arthritis, showed that 63 percent of subjects taking a ginger extract reported pain relief, compared with 50 percent of those taking placebo. Those taking ginger also reported less pain after walking 50 feet. Although this constitutes preliminary evidence, more research needs to be done to confirm it.

A study in rats showed ginger to have anti-inflammatory effects, reducing swelling (Sharma et al., 1994). No studies to date have investigated this effect in humans. Ginger has fever-reducing effects in rats, but this also needs to be tested in humans.

Ginger is known to reduce blood clotting in a manner similar to other NSAIDs. This effect has been demonstrated in humans who took oral doses of ginger. However, fresh ginger had this effect, but dried ginger did not. It's unlikely that this represents a real danger to people taking blood thinners (anticoagulant medications such as aspirin and warfarin), since ginger is usually consumed dried or cooked.

CHILI PEPPERS

It seems ironic to discuss **chili** peppers in a chapter about painkillers, since chili peppers are so well known for *causing* pain. But used in the right way, they actually can act as an analgesic. Chili, or cayenne, peppers are characteristically known for the burning sensation they produce on contact. Among the many species of chili peppers, the most commonly known are *Capsicum annuum, C. frutescens, C. baccatum, C. chinense,* and *C. pubescens* (see Figure 8.4). They are eaten on a daily basis by one-quarter of the world's population. The many names by which we know chili peppers refer to the different varieties, which differ by their size, color, shape, texture, and hotness (see Table 8.5). As a general rule, the smaller the pepper, the hotter. The name *chili* is used here as a blanket term for all hot peppers.

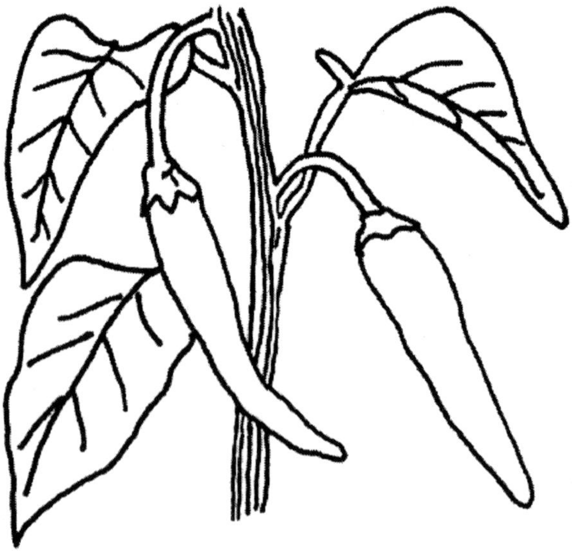

FIGURE 8.4. Chili pepper *(Capsicum annuum).* Copyright 2004 by Marcello Spinella.

TABLE 8.5. Degree of heat in chili peppers.

Class	Scoville units	Examples
Sweet	0	Bell pepper
Mild	1,000-3,000	Ancho or poblano peppers, ortega chilies, paprika
Medium	3,000-6,000	Jalapenos, rocotillo, hot banana, Christmas
Hot	5,000-100,000	Cayenne, tabasco, serrano
Very hot	80,000-300,000+	Chiltepin, habanero, rocoto,Thai

Chili peppers are a hollow, boxlike fruit. The name *Capsicum* comes from the Latin word *capsa*, meaning box. *Chili* is the name for the plant in Nahuatl, the language spoken by the Aztecs. The fruit contains numerous disc-shaped seeds, which are hot because of their proximity to the ribs of the pepper—the hottest part. Chilies are native to central America, but were brought to Europe by the Spanish, and introduced to Asia and Africa by the Portuguese. They are now grown in many areas with warmer climates.

It is interesting that most of our modern uses of chili peppers were thought of by ancient people of Central and South America. We use an extract of chili peppers to make a painful, irritating self-defense spray. In the same manner, the Incas burned chilies and fanned the smoke toward their enemies. Just as we now use the active chemical in chili for analgesia, Native Americans rubbed chili on the gums to relieve the pain of a toothache. Similarly, eunuchs serving Chinese emperors were castrated after hot pepper extracts were applied over a long period to their scrotums.

Active Chemical in Chili

The main chemical in chilies producing the burning sensation is called **capsaicin.** It was extracted from chilies and named by Thomas Thresh in 1846. In 1912, Wilbur Scoville developed a scale to rate the hotness of chilies using Scoville units. A pepper with 1,000 Scoville units means that an extract must be diluted 1:1,000 before it ceases to be hot. So 1 ml of extract must be put into 1,000 ml (1 l) of water before you can no longer detect its hotness. Some of the hotter varieties of chilies (e.g., the habanero) can reach an amazing 300,000 Scoville units. Pure capsaicin rates an astonishing 16 million Scoville units.

How Chili Peppers Work

The capsaicin in chili peppers attaches to sensory receptors in the skin called **vanilloid receptors.** These receptors can be stimulated by both chemical (i.e., capsaicin) means or heat. Thus, chili peppers trick a type of sensor into reporting the sensation of heat, even though the temperature of the skin may be normal. As with any type of stimulation that becomes intense enough, pain nerves are eventually activated and pain is experienced. As with most forms of pain, this causes a release of other neurotransmitters in the spinal cord and brain that further transmit the pain signals. One such neurotransmitter is called substance P. So, when hot pizza activates these nerves, it's because actual heat is burning the inside of your mouth. When chili peppers cause a burning feeling, it's because they're chemically tricking your pain- and heat-sensing nerves.

Effects of Chili Peppers

Capsaicin may affect pain in a few different ways. In the case of sore muscles, from exercising for example, it may speed healing by increasing blood flow to the area. Chili peppers may also cause counterirritation, reducing pain by causing nearby irritation. This is not well understood, but it's why people tend to scratch painful cuts.

Capsaicin is available at pharmacies as a topical cream that causes an immediate burning feeling when applied to the skin. However, when applied daily over a period of time, the burning and any other pain in the region diminishes or disappears. It seems that the constant presence of capsaicin makes pain nerves work overtime, releasing their neurotransmitter substance P. Substance P is used up faster than it is made, in a sense, and the nerves become desensitized to pain.

The cream is applied directly to the part of the body where the pain is occurring. It has been used for many conditions causing pain, such as arthritis (both rheumatoid and osteoarthritis), nerve dysfunction (e.g., postherpetic neuralgia, diabetic neuropathy) and amputation (e.g., postmastectomy and phantom limb pain). Unfortunately, a little paradox makes it impossible to do a true controlled study of capsaicin for pain. In a controlled study, the drug must be compared to a placebo that is indistinguishable from the actual drug. Since capsaicin causes burning, it would be easy for research subjects to tell the difference between the active cream and the inactive placebo cream. On the other hand, if we developed a burning placebo, then it would probably have the same effect as the capsaicin, in which case it wouldn't be inactive, as a placebo is supposed to be.

Side Effects of Chili

The main side effect of using a capsaicin cream is the unpleasant burning sensation. Some people find this intolerable, and as many as 30 percent of subjects drop out of research studies. Using too much can actually cause blisters and ulcers in the skin. Even though the capsaicin isn't truly burning the skin, the skin reacts as if it were. Eating too many chili peppers or ones that are too hot causes pain in the digestive tract and diarrhea. Although chilies *feel* as though they're burning, they actually lower body temperature when given to animals, but it has yet to be shown that this happens in humans. If so, it may be why chilies are so popular in tropical climates: to keep people cool.

Once the applied cream dries up, the skin area should be carefully washed. Capsaicin remaining on the skin can disperse into the air, and cause lung irritation and coughing.

STIMULANT ANALGESICS

As mentioned in Chapter 4, there are three main groups of stimulants. The herbs within each group work in similar ways and have similar psychological effects. These groups are the caffeine-containing herbs, tobacco-like herbs, and amphetamine-like herbs. In addition to being stimulants, they also have painkilling effects—either by blocking pain transmission or by activating the analgesia system in the brain.

Caffeine-Containing Analgesic Herbs

Coffee, tea, chocolate, guarana, maté, and cola are all plants that contain caffeine (and the closely related drugs theobromine and theophylline). Their stimulant effects are due to their ability to block the neurotransmitter adenosine, which activates the brain. However, adenosine also transmits pain messages. So when caffeine blocks adenosine, it is blocking pain transmission.

Caffeine alone may produce analgesic effects in some forms of pain, but clinically it is most often used in combination with other painkillers. Caffeine enhances the painkilling effects of NSAID drugs such as acetaminophen and aspirin, also speeding the onset of action. It's found in common over-the-counter brands such as Excedrin. Caffeine also increases the analgesic effects of opioid drugs such as morphine. This is a convenient effect since opioids can also cause unwanted sedation, which is counteracted by caffeine.

Caffeine alleviates migraine headaches because it constricts blood vessels. Caution should be exercised, however; regularly using caffeine and suddenly stopping it can cause rebound migraines in people who suffer from them.

Tobacco-Like Analgesic Herbs

Three tobacco-like stimulants are discussed in Chapter 4: tobacco, lobelia, and areca. In addition to having stimulant effects, they are also painkillers. This is because they stimulate receptors for the neurotransmitter acetylcholine, which are involved in pain inhibition as well as alertness. The drugs in these plants (mainly nicotine in tobacco, lobeline in lobelia, and arecoline in areca) imitate acetylcholine and activate the pain inhibition system. Although smoking tobacco may not be prescribed, it is not uncommon to see smokers seeking out a cigarette when they are in pain.

Amphetamine-Like Analgesic Herbs

Coca

Cocaine's stimulant actions (prolonging the actions of serotonin and norepinephrine) also activate the analgesia system and reduce pain. Cocaine's painkiller effects have been shown in animals. Cocaine also boosts the painkilling effects of opioids such as morphine. Of course, the strong stimulant effects and addiction potential of cocaine make it of very limited usefulness in treating pain.

Khat

Khat similarly works by boosting the actions of serotonin and norepinephrine, activating the analgesia system. This is accomplished by cathinone, the main active drug in khat. Animal studies of cathinone show that it has analgesic effects. As with cocaine, the strong stimulant effects and addiction potential of khat limit its usefulness in treating pain.

Ephedra

Ephedrine, the main drug in ephedra (or ma huang), imitates norepinephrine and activates its receptors. So one would expect it to have painkilling effects by activating the analgesia system. Ephedra doesn't seem to have been tested by itself for analgesic effects, but it has been looked at in combination with other painkillers such as morphine. Ephedrine does boost the painkilling effect of morphine: in one study, giving morphine plus ephedrine was equivalent to doubling the dose of morphine. As with caffeine, this could be a convenient combination since, in addition to boosting the analgesic effect, the stimulant effects of ephedra would counteract the sedative effects of morphine.

Cannabis for Analgesia

Cannabis (marijuana) is discussed in much greater detail in Chapter 10. However, it is mentioned briefly here for its analgesic effects. THC, the main active constituent of cannabis, is an effective analgesic drug, even when given by itself. THC attaches to receptors in the nervous system and activates the analgesia system. THC also enhances the painkilling effects of other drugs such as morphine when given in combination. In fact, they can produce a cross-tolerance: using morphine over time will cause a tolerance to both morphine's and THC's analgesic effects (and vice versa). This suggests that the two drugs both work through somewhat similar systems in the brain. Likewise, THC-like chemicals made in the brain (called endocannabinoids) play a natural role in controlling pain.

Analgesic Effects of THC in Humans

Despite ample evidence from animal studies that THC has analgesic effects, only a few controlled studies have attempted to prove this in humans. Studies of people with cancer pain show that doses of 10 to 20 mg of THC (taken orally) reduce pain. The 10 mg dose of THC was roughly equivalent to 60 mg of the standard painkiller codeine. This dose of THC produced some sedation, but no other psychological effects. Although the 20 mg dose was more effective, it also caused more side effects such as dizziness, unsteadiness, blurred vision, and psychological effects. A more recent study has shown that smoked marijuana is also a painkiller. This study used healthy volunteers and showed that it increased their tolerance for pain.

Although these studies provide solid evidence for the painkilling effects of cannabis, more clinical research is needed to firmly establish it for this use. Regardless, the tentative legal status of cannabis has hindered proper research and clinical use.

LOCAL ANESTHETIC HERBS

Local anesthesic drugs block nerve transmission and stop all sensation, resulting in temporary numbness. Anyone who has had a tooth drilled by the dentist is familiar with this feeling (or lack of feeling). At the same time, these drugs affect motor nerves as well, causing temporary muscle weakness. These drugs are used for pain prevention, often injected or given in a topical ointment applied to the region of the body in question. Some common pharmaceutical local anesthetics include Novocain and Xylocaine. The herbal drugs in this category are coca, clove, allspice, and kava (see Table 8.6).

Coca

Cocaine, from the coca plant, is the first drug that gained widespread medical use for its local anesthetic effects. This effect was long known to the natives of Peru, who noticed that chewing coca leaves caused numbness in the mouth. In 1860, Albert Nieman isolated cocaine from the coca plant and named it, and Carl Koller started using cocaine to numb the eye for surgery in 1884. The strong

TABLE 8.6. Herbal local anesthetics.

Herb	Active chemical
Coca	Cocaine
Clove	Eugenol
Allspice	Eugenol
Kava	Kavalactones

stimulant effects of cocaine can make it impractical for many medical uses, so similar drugs were invented that have less stimulant effect. For example, the drug procaine (Novocain) was created in 1905.

How Cocaine Works

Cocaine blocks electrical transmission in nerves, so that pain messages are stopped in the body and never reach the spinal cord and brain. Cocaine and similar drugs stop this electrical transmission by blocking sodium channels on the surface of neurons. This prevents the electrical signal from spreading along the nerve. All local anesthetic drugs seem to work in this manner.

Local anesthetic drugs such as cocaine cause something called use-dependent block. Essentially, this means that the more active a nerve is, the more blocking effect the drug has. Nerves that are relatively inactive are not affected much, but highly active nerves are blocked more heavily. Also, smaller nerves are more sensitive to local anesthetics than large-diameter nerves. As a result, nerves for pain and touch, which are smaller in diameter, are affected more than nerves controlling muscles, which are larger. This means that local anesthetics conveniently reduce pain without causing muscle weakness.

Cocaine is probably the only known plant that has both analgesic and anesthetic effects. Most drugs do one or the other. This may seem confusing at first, but these two different effects are caused by the two distinct actions of cocaine. In the brain, cocaine boosts the neurotransmitters serotonin and norepinephrine, which causes the painkilling (analgesic) effect. But the local anesthetic effects work on nerves in the body, by blocking their electrical signals.

Clove

Clove comes from the tree *Syzygium aromaticum* (see Figure 8.5). It is an evergreen tree that grows up to 65 feet in height. The cloves used as a spice for cooking come from the pink flower buds, which turn reddish-brown after drying. Clove oil, used medicinally for local anesthetic effects, is steam-distilled from the flowers. Clove is sometimes ground and added to tobacco to produce clove cigarettes.

Active Chemical in Clove and How It Works

The main anesthetic chemical in clove is called **eugenol.** Eugenol seems to work by blocking electrical transmission in nerves, probably in a manner similar to cocaine and the pharmaceutical local anesthetics. Clove oil is often used on the gums for toothaches and teething babies. It has also been mixed with zinc oxide and used by dentists as a temporary tooth filling.

How eugenol might affect the brain when a person smokes clove cigarettes is uncertain. Pharmaceutical local anesthetics must some-

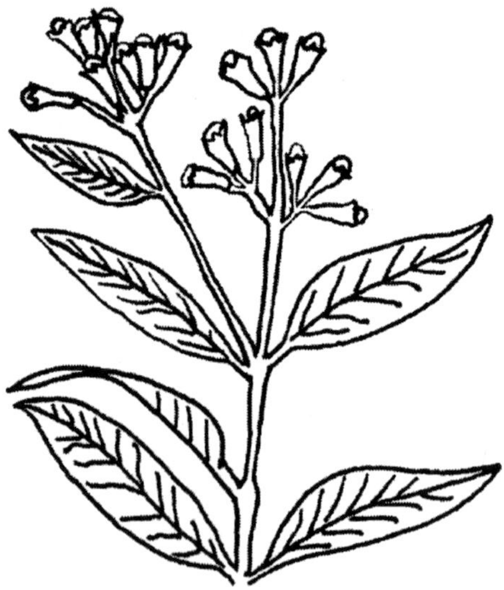

FIGURE 8.5. Clove *(Syzygium aromaticum).* Copyright 2004 by Marcello Spinella.

times be used intravenously to treat abnormal heart rhythms. When given this way, they reach the brain and cause emotional effects such as depression and anxiety. Whether the eugenol in clove cigarettes is strong enough to reach the brain and create psychological effects is uncertain. People who smoke clove cigarettes do not seem to notice much of a difference between clove cigarettes and regular tobacco cigarettes, other than clove's distinct smell.

Side Effects of Clove

Clove oil is listed as generally regarded as safe by the U.S. Food and Drug Administration when used properly. Clove oil is very concentrated, so it should only be used topically, and it can irritate sensitive skin. When used for toothaches and teething babies, it is rubbed on the gums (as with a cotton swab) but should not be swallowed. Swallowing clove oil can cause nausea, and large overdoses can have more serious effects. One case was reported of a child who drank clove oil, which caused a coma, seizures, abnormal blood clotting, and liver damage. As with any other potent medication, clove oil should be treated with caution, especially around children.

Since clove cigarettes also contain tobacco, they carry all of the dangers of regular tobacco smoking (i.e., cancer, emphysema, and heart disease). Clove cigarettes also seem to be harder on the lungs than regular tobacco cigarettes. The eugenol in clove anesthetizes the breathing passages, blocking the gag reflex and making it harder to cough up fluids. Medical cases have been reported of swelling in the lungs (pulmonary edema), pneumonia, and bronchitis in association with clove cigarettes.

Allspice

Allspice *(Pimenta dioica)* is another culinary spice that has local anesthetic effects. It's an evergreen tree native to Central and South America. Historically in Guatemala, crushed allspice berries were applied to bruises, joints, and muscles for pain. As with clove, allspice contains eugenol, which blocks electrical nerve transmission. However, allspice has lower amounts of eugenol than clove, so dentists have historically favored clove oil.

Allspice is listed by the U.S. Food and Drug Administration as generally regarded as safe when used appropriately. As with clove oil, allspice oil should only be used topically and it may irritate sensitive skin. Since it contains eugenol, swallowing it could cause nausea and potentially more serious effects such as coma, seizures, liver damage, and abnormal bleeding.

Kava

Kava is a sedative herb (discussed in Chapter 6) that also has local anesthetic effects. The active chemicals in kava, the kavalactones, also block electrical transmission in nerves. Touching the tip of the tongue to kava extract will make it go numb, in addition to the strongly bitter and astringent taste. Kava hasn't been used much as a local anesthetic, although this effect is widely known. However, in recent years, some herbal preparations (teas, lozenges) have reached the market that use kava to soothe sore throats.

Chapter 9

Herbal Hallucinogens

Of all the psychoactive medicines that come from plants, none have so captured our attention as the **hallucinogens.** They have a powerful ability to alter our perceptions, mood, and thoughts—in short, to alter our very sense of reality. Not surprisingly, hallucinogenic plants have often been used for religious and spiritual ceremonies. Modern science has been no less fascinated by these plants for what they may teach us about the fundamental workings of the mind and brain.

Although a few hallucinogenic drugs have been invented by humans (e.g., PCP or angel dust, ketamine or Special K), the vast majority of them come from plants. They're not even confined to any certain type of plant, but rather are found in virtually every type known: shrubs, vines, trees, leafy plants, fungi, grasses, and cacti. Most grow on land, but some are aquatic plants, such as the blue water lily. They're even spread out geographically: hallucinogenic plants can be found in all of the continents except Antarctica. Over a hundred known hallucinogenic plants could easily be named, and there are probably many more we don't yet know about.

The use of hallucinogens is fairly common in Western cultures. A survey done in 1990 by the National Institute on Drug Abuse suggested that 7.6 percent of the American population over age twelve had tried a hallucinogenic drug at least once in their lives, and 1.1 percent had done so in the previous twelve months. Hallucinogens are most commonly used by people between the ages of eighteen and twenty-five years.

WHAT TO CALL THEM?

There has been much discussion of what to call this class of drugs. Psychoactive drugs are most often classified according to their most prominent effects (as they are in this book). A toxicologist named Louis Lewin came up with the name "phantasticants" in 1924. Although the name is appropriate, it's rarely, if ever, used nowadays. The name "psychedelic" translates from Greek to mean "mind manifesting" or "mind expanding," but it's not clear exactly what mind expansion means. The American psychologist William James (1902) suggested that "our normal waking consciousness . . . is but one special type of consciousness, whilst all about it, parted from it by the flimsiest of screens, there lie potential forms of consciousness entirely different" (p. 378). In this sense, hallucinogens show us just how much of what we call "reality" is really something that is put together by our brains. A real, objective world exists, but all that we know of it is only what our brains can piece together. As a case in point, hallucinogenic drugs distort our reality: reality doesn't change, but our perception of it does. Not all hallucinogens really do this, however. Some create more of a confused, delirious state rather than what one would consider mind expansion.

Sometimes the term **psychotomimetic** is used, drawing comparisons between the effects of hallucinogens and the symptoms of schizophrenia and other psychoses. However, the resemblance is only superficial. While drug-induced hallucinations tend to be more visual, the hallucinations of schizophrenia are most commonly auditory (e.g., hearing voices). Schizophrenia also has many other symptoms (e.g., cognitive deficits, social withdrawal) that are not caused by most hallucinogenic drugs.

Hallucinogen comes from the Latin *alucinari*, meaning to "wander in mind." Literally, *hallucinogen* means "an agent that causes hallucinations," but it is not the only kind of drug that can cause hallucinations. Very high doses of other drugs (e.g., cocaine, opioids, and cannabis) can do the same. So the term *hallucinogen* is reserved for those drugs for whom hallucinations are a very prominent effect, even at low doses. Strictly speaking, a hallucination is a false perception: seeing, hearing, smelling, tasting, or touching something that is not really there. However, hallucinogenic drugs affect more than our sensory perceptions; they also strongly alter mood and thinking.

It seems there is no completely satisfactory term for this class of drugs. None can completely cover all of the effects of these drugs. *Hallucinogen* is perhaps the best fit and is used throughout this chapter.

THE HISTORICAL USES OF HALLUCINOGENIC PLANTS

Since hallucinogenic plants have such a powerful ability to alter our sense of reality, it is no surprise that early humans universally regarded them as divine plants. Hallucinogenic plants occur throughout the world, and the human cultures that have found them have almost always treated them as spiritual and magical tools. Many ancient cultures incorporated them into religious ceremonies.

We now understand that hallucinogenic drugs create their effects by altering activity in the brain. Our understanding of biology and the brain has progressed to the point where this is the most sensible conclusion. However, before this type of understanding existed, it is easy to see how supernatural explanations were common.

Modern interest in the hallucinogens greatly increased after the creation of **LSD,** or *d*-lysergic acid diethylamide, by Albert Hofmann in 1949. Hofmann was a chemist working for the pharmaceutical company Sandoz in Switzerland. He inadvertently ingested some of the drug and later reported his experiences. The writer Aldous Huxley experimented with mescaline and described these experiences in his book *The Doors of Perception,* published in 1954. Popular experimentation with hallucinogens blossomed during the 1960s in the West. The experiences of people such as Ken Kesey were described in *The Electric Kool-Aid Acid Test* by Tom Wolfe. Ironically, some of the first exposure that Americans had to hallucinogens was as volunteers in government experiments: Ken Kesey volunteered for such experiments in 1959 at Menlo Park Veterans Hospital.

While some experimented with hallucinogens in that era in an attempt to expand consciousness, others proposed that they could create a temporary schizophrenia in humans. This and countercultural dissent contributed to the placement of hallucinogenic drugs on Schedule I of the Controlled Substances Act of 1970. This restricted their legal use, claiming that they have a high potential for abuse and lack therapeutic value. In 1993, the Religious Freedom Restoration Act allowed members of the Native American Church to legally use peyote.

Hallucinogens may be classified in many different ways. The most illustrative is to group them according to how they work in the brain. Like other psychoactive drugs, hallucinogens with similar mechanisms have similar psychological effects. In this book, they are grouped into three classes: LSD-like hallucinogens, nightshade hallucinogens, and amanita mushrooms, which are in a class by themselves.

LSD-LIKE HALLUCINOGENS

The LSD-like hallucinogens are a diverse group of plants, consisting of mushrooms, cacti, vines, and trees. What they all have in common is that they contain chemicals that seem to interact with the serotonin system of the brain.

Ergot Fungus

The **ergot** fungus *(Claviceps purpurea)* grows on rye and other grains, particularly during wet seasons (see Figure 9.1). It is a hard black fungus that protrudes from the grains. Insects carry the spores to plants, where they grow into the grain and rob it of nutrients.

The poisonous effects of ergot were known to the Mesopotamian cultures of Assyria and Babylonia. An Assyrian cuneiform tablet dat-

FIGURE 9.1. Ergot *(Claviceps purpurea)*. Copyright 2004 by Marcello Spinella.

ing from around 600 B.C. refers to ergot as a "noxious pustule in the ear of grain." Ironically, it seems that it wasn't until 1676 A.D. in Europe that the connection was made between the ergot fungus itself and the toxic effects of eating ergot-infected grain. These toxic effects are referred to as ergotism.

Ancient Romans called ergotism *ignis sacer,* or holy fire. This name was retained through the European Middle Ages, and refers to the burning feeling ergotism causes in the extremities. Ergotism was no small matter, and casualties from it were significant: one epidemic in Europe in 944 A.D. caused the deaths of 40,000 people.

Ergotism also became known as St. Anthony's fire. St. Anthony was a monk who lived in the third and fourth centuries A.D. In 1039 A.D., a shrine was named after him where an outbreak of ergotism occurred, and it became a popular object of pilgrimage for people suffering from ergotism. A pilgrimage was probably an effective way to cure ergotism: by leaving home, people ate foods other than their ergot-infected bread, and the walking probably improved blood circulation, which is reduced by ergot.

Cold, damp growing conditions promote the growth of ergot, and modern outbreaks of ergotism still occur, such as ones in the Ukraine and Ireland in 1929, and in Belgium in 1953. Although it cannot be proven, it has been suggested that the Salem witch trials in 1692 may have been partly sparked by ergotism.

Ergotism

People who eat ergot fungus experience many horrible symptoms, which are due to nonhallucinogenic chemicals in the fungus. Ergotism takes two general forms: (1) the neurological form, which consists of seizures and hallucinations; and (2) the vascular form, which causes gangrene of the extremities. Ergot chemicals constrict blood vessels until there is no longer any blood flow. As a result, noses, earlobes, fingers, toes, and even feet were known to fall off.

Active Chemicals

The two main hallucinogenic chemicals in the ergot fungus are *d*-isolysergic acid and *d*-lysergic acid. The famously known hallucinogen LSD (*d*-lysergic acid diethylamide) is a synthetic drug derived from those naturally occurring in ergot. All three are hallucinogenic, but LSD is more potent due to its chemical modifications. The majority of research in this area focuses on LSD, but the results apply to most similarly acting drugs.

LSD

LSD is odorless and tasteless and is effective in microgram doses. Doses of 25 to 50 μg (i.e., millionths of a gram) cause an intensification of perceptions, while larger doses are needed for full hallucinations. The effects of the drug typically last six to twelve hours. Taking LSD daily for a few days causes tolerance to develop, so the same dose taken repeatedly becomes ineffective. LSD causes cross-tolerance with other similar hallucinogens, such as mescaline and psilocybin, suggesting that they all work the same way in the brain.

How LSD Works

LSD works by stimulating the serotonin system of the brain. More specifically, it stimulates a particular serotonin receptor called the 5-HT_2 receptor. This seems to be the common mechanism of all LSD-like hallucinogens. The potency of all such hallucinogens relates directly to how strongly they attach to this receptor in the brain. Taking LSD repeatedly causes the 5-HT_2 receptor to lose sensitivity, which explains why taking it every day makes it less and less effective.

Autonomic Effects

LSD activates the sympathetic nervous system, which increases heart rate and blood pressure, and dilates the pupils.

Brain Imaging

Studies that look at the electrical effects of LSD show that it activates the brain. This makes sense in light of the intense hallucinatory experiences that LSD can trigger. This activation appears to occur through LSD's effects on the serotonin system of the brain.

Psychological Effects

The effects of LSD and their intensity are a matter of how much drug is taken. It causes an intensified awareness of sensations, so that ordinary, mundane sensations become vivid. Whereas lower doses cause illusions (distorted perceptions of real objects), higher doses cause hallucinations (perceiving something that isn't really there). Some of the common distortions include intensification of colors and alterations of shading, movement, and shape. Other visual illusions include macropsia or micropsia, which are the perceptions that objects are getting larger or smaller, respectively (think of Alice in Wonderland). Afterimages may trail after objects. Higher doses can cause one to see geometric patterns and eventually fully formed hallucinatory experiences. In some instances, people may vividly reexperience past experiences. The hallucinations are most typically visual, beginning two to three hours after taking the drug. Auditory hallucinations (i.e., hearing things) are less common. However, altered perceptions of bodily sensations do occur (e.g., feeling as if one is floating). The passage of time also seems slow to the LSD user, so that minutes may seem like hours.

LSD affects not only sensory perceptions, but also emotion. The emotional meaning of things becomes intensified, so that experiences seem profound or magical. LSD experiences often have a religious or mystical quality, which is consistent with the spiritual uses of hallucinogens by so many different cultures. People report feelings of unity, whereby their personal boundaries diminish and they feel a part of the larger world around them. A feeling of transcending time and space also occurs, and a sense of awe and sacredness is present. These religious feelings are typically accompanied by positive mood, often described as overwhelming joy and tranquility. Finally, a sense of ineffability accompanies these experiences; they may be described but they cannot really be conveyed with words.

Not all LSD experiences are as positive as these religious experiences. They vary greatly according to one's frame of mind and the setting (the situation in which the drug is taken). People taking the drug alone are more likely to experience anxiety, while those taking it in a group setting are more likely to have a positive experience. A "bad trip" is essentially an LSD experience that has become unpleasant and anxiety provoking. Instead of feeling like part of the universe, users may just feel detached from reality and even their own bodies, which can be a frightening experience.

Addiction

LSD does not cause addiction. It is not strongly reinforcing the way drugs such as cocaine, heroin, and nicotine are. Neither laboratory animals nor humans tend to take it in a compulsive, repetitive pattern as they do drugs of abuse. As a result, nobody ever shows up at an addiction clinic because they cannot stop using LSD. This is not to say that it's impossible to misuse LSD. The strong distortions and experiences that it causes make a person unable to safely perform most everyday tasks, such as driving a car or using a stove. Sensible use of hallucinogens would most likely follow the ceremonial use exemplified by the Native American Church: respectful and infrequent use in the context of a positive, social setting.

Side Effects

Massive overdoses of LSD are needed to have lethal effects. Fatal doses in rodents are astronomically high and no human fatalities have ever been reported from an overdose of LSD. However, a case was reported in which eight people snorted pure LSD, mistaking it for cocaine. It was estimated that they each consumed between 10,000 and 100,000 μg (remember that even 100 μg will cause hallucinations). They experienced severe confusion, hallucinations, and bleeding (such high doses inhibit blood clotting), but all survived and recovered.

Bad Trips

As mentioned, a bad trip is essentially a panic reaction during a hallucinogenic experience. LSD also slows the perceived passage of time, so the unpleasant experience may seem as though it may never end, further feeding the panic. Bad trips seem to respond best to supportiveness and being in a familiar environment, although some argue that it can worsen the condition in some cases. Anti-anxiety and antipsychotic drugs can be used in these cases.

Flashbacks

Flashbacks are formally called **hallucinogen-persisting perception disorder** (or HPPD). They are instances when a person later reexperiences parts of the hallucination. Certain factors are known to precipitate flashbacks, such as use of cannabis, anxiety, fatigue, and sudden movement into a dark environment. In a few cases, they have been triggered paradoxically by the antipsychotic drug Risperdal. Flashbacks may be a normal function of memory that are exaggerated because of the intense nature of the hallucinogenic experience.

Half of LSD users who have flashbacks experience them five years after using the drug. Most commonly, people see shapes, fleeting objects in their peripheral vision, flashes of color, and afterimages.

Other Side Effects

LSD does not pass through the placenta, but it can cause uterine contractions and spontaneous abortions. Use during pregnancy should be avoided. Some reported in the 1970s that LSD could damage DNA, but this only seems to happen at unrealistically high doses, if at all.

Psilocybe Mushrooms

Several species of mushrooms have hallucinogenic effects. Many are in the *Psilocybe* genus, including *Psilocybe aztecorum, P. caerulescens,* and *P. mexicana*. Another related species is *P. cubensis.*

Many Native American civilizations in Central America knew about the effects of these mushrooms. The Aztec called them *teonanacatl,* meaning "divine flesh," and reserved their use for religious ceremonies. The religious use of hallucinogenic mushrooms was also practiced by the Mazatec, Chinantec, Mije, Zapotec, and Mixtec of Oaxaca. The Mazatec still actively employ mushrooms in religious ceremonies, referring to them as *Nti-si-tho*. The Maya may also have used mushrooms ceremonially.

Active Chemicals

The main hallucinogenic chemicals in these mushrooms are called **psilocybin** and **psilocin.** They work similarly to LSD, by stimulating the serotonin system of the brain. They are both approximately 1/200th as potent as LSD.

Effects of Psilocybe Mushrooms

The effects of **psilocybe** mushrooms are very similar, if not identical, to those of LSD. Many people experience nausea and vomiting after eating these mushrooms. They are also reputed to have a horrible taste. However, the effects of psilocybe mushrooms are considerably shorter, lasting four to six hours compared to LSD's six to twelve hours.

Peyote

Peyote *(Lophophora williamsii)* is a small, spineless cactus that grows in the deserts of Mexico and the Rio Grande valley (see Figure 9.2). The peyote cactus is prepared by cutting the top, or crown, of the plant and drying it into buttons. The buttons are then eaten or made into a tea.

FIGURE 9.2. Peyote *(Lophophora williamsii).* Copyright 2004 by Marcello Spinella.

The name *peyote* derives from the Nahuatl word for the cactus, *peyotl*. Early European reports date peyote at least as far back as 300 B.C. among the Chichimecs and Toltecs of ancient Mexico, but samples found in caves have been dated to 8,000 years ago. The Spaniards tried to eradicate the use of peyote, but instead only drove it into secrecy. Its use began to spread among tribes in the United States through the late 1800s, and the Native American Church was founded in 1918. It is now estimated to have 250,000 members.

Mescal Bean

Neither the peyote cactus nor mescaline should be confused with the mescal bean *(Sophora secundiflora)*. They are sometimes confused because of the similarity of the names, but the mescal bean comes from an entirely different plant. It also has psychoactive effects and was used by the Arapaho and Iowa tribes in North America.

Active Chemicals

More than fifty-five active chemicals have been isolated from peyote. **Mescaline** is the main drug, and by itself will produce the full hallucinogenic effects. The effects of the drug last approximately ten hours. Mescaline is not broken down by the liver, but instead it is excreted unchanged in urine. Although mescaline is commonly known, it is rare and expensive. Most samples of supposed mescaline bought illegally are simply LSD.

How Mescaline Works

Mescaline most likely works in a manner similar to LSD, by stimulating the serotonin system of the brain. Ounce for ounce it is less potent, so doses of 200 to 500 mg are needed to create psychedelic effects.

Effects of Mescaline

Little is known about the effects of mescaline compared to those of LSD. However, when people are given equally potent doses of LSD and mescaline, the effects are so similar that they cannot be told apart, further supporting the idea that these drugs probably work in the same manner.

Brain imaging studies of people taking mescaline show that it activates the brain, similarly to LSD. This might be expected of a drug that causes such intense perceptions and experiences. One study showed that it increases blood flow particularly to the right side of the brain, which is more involved in aspects of emotion.

Side Effects

A common side effect of peyote, as with several other hallucinogens, is nausea. There is no evidence that mescaline causes genetic damage. The Huichol Indians have a 1,600-year tradition of using peyote in religious ceremonies, but show no evidence of chromosomal damage. Mescaline is teratogenic and, like all hallucinogens, should not be used during pregnancy.

Other Hallucinogenic Cacti

Other species of cacti also have hallucinogenic effects. The San Pedro cactus *(Trichocereus pachanoi),* which grows in the Andes in South America, is a tall, columnar cactus that contains mescaline. Cawe *(Pachycereus pecten-aboriginum)* is another hallucinogenic cactus that grows in Mexico.

Yage

Yage (also known as yaje, hoasca, or ayahuasca) is a mix of hallucinogenic plants growing in the Amazon, of which *Banisteriopsis caapi* and *Psychotria viridis* are common ingredients. *Banisteriopsis* is a liana (thick, woody vine), and *Psychotria* is a shrub or small tree. The combination of plants is reported to prolong and intensify the hallucinogenic effects. The American writer William S. Burroughs popularized the drug in his letters to Allen Ginsberg, later published as *The Yage Letters*.

Active Chemicals

The main hallucinogenic ingredients of *Banisteriopsis* are a family of chemicals called **beta-carbolines,** including harmine and harmaline.

Psychotria and some species of *Banisteriopsis* also contain a hallucinogenic chemical called dimethyltryptamine (more commonly called DMT).

How Yage Works

Both DMT and the beta-carbolines seem to work similarly to LSD, by stimulating the serotonin system of the brain. Interestingly, the brain seems to produce its own beta-carbolines, since harmine has been found to occur naturally in the human brain. What function this chemical normally serves is unknown. DMT taken by mouth is not well absorbed into the body. Accordingly, it is usually snuffed or smoked, or mixed with other plants that improve its absorption.

Effects of Yage

The effects of yage are very much like those of the other LSD-like hallucinogens. Some users have reported that unlike other hallucinogens, ecstatic feelings do not occur with yage. Amazonian Indians taking yage often report seeing snakes and jaguars. This is probably not due to the drug itself, but rather because those animals are important symbols in Amazonian cultures. One thing that distinguishes DMT is that it has very brief effects. The effects come on very quickly and end in about thirty minutes.

Side Effects

Beta-carbolines inhibit the enzyme monoamine oxidase, which could potentially interfere with antidepressant drugs and cause serious side effects. Harmaline crosses the placental barrier and should not be used by pregnant women. Harmaline also has toxic effects in laboratory cultures of neurons. Whether this effect is strong enough to occur when taken by humans has never been researched. Neurological symptoms have not been noted in people who have taken the drug.

Morning Glory and Related Plants

Several species of flowering vines have hallucinogenic effects. These are commonly referred to as **morning glory.** The most popular is probably ololiuqui *(Turbina corymbosa)*. Ololiuqui is the Aztec word for the seed, literally meaning "round object." Its hallucinogenic effects were known to the Aztecs, who used it in religious ceremonies. Another species was known to the Zapotec as *badoh negro (Ipomoea violacea)*. These flowers are commonly available in garden shops in varieties called 'Heavenly Blue,' 'Pearly Gates,' and 'Flying Saucers.' The Hawaiian baby woodrose *(Agyreia nervosa)* is another related flowering plant species that contains hallucinogenic chemicals.

Active Chemicals

The identified psychoactive constituents of the morning glories are relatives of lysergic acid (e.g., lysergic acid amide, lysergol, lysergic acid hydroxyethylamide, and isolysergic acid amide). As the names imply, these are closely related to LSD (lysergic acid diethylamide). However, LSD is a chemically altered version that is 100 times more potent, ounce for ounce.

How Morning Glory Works

Since the active chemicals in morning glory are so similar to LSD, they probably work through similar if not identical means: stimulating the serotonin system. They are much less potent, so higher doses are needed to cause hallucinations compared to LSD, which is effective in millionths of a gram. Side effects, such as nausea and vomiting, are more common with morning glory.

Nutmeg and Mace

Most people do not know that the common cooking spices **nutmeg** and **mace** are hallucinogenic. Both of these spices come from the same tree, *Myristica fragrans*. It is an evergreen tree, native to Indonesia but now grown in many tropical regions, growing up to 15 meters in height, bearing flowers and a fleshy, round fruit. Its use in Europe began via trade with Arabs in the first century A.D. However, nutmeg and mace were well known in ancient Egypt and India, where nutmeg is sometimes snuffed together with tobacco.

Nutmeg is made from the seed of the tree, while mace is prepared from the fleshy covering that surrounds the seed. The psychoactive chemicals in nutmeg and mace are **myristicin** and **elemicin.** These two are similar in their chemical structure to the drug mescaline.

A large amount of spice must be consumed (about 15 g) to produce psychoactive effects. Physical effects are prominent, and include rapid heartbeat, dry mouth, and flushing of the skin. The hallucinogenic and emotional effects begin after two to five hours, and can include anxiety and an unpleasant detachment from reality. A hangover-like feeling is also reported to occur afterward. It has been rumored that nutmeg and mace are used in place of illegal drugs in prisons, but illegal drugs can be obtained virtually anywhere. Further, the side effects of nutmeg and mace are negative enough to usually discourage their use.

Iboga

Iboga *(Tabernanthe iboga)* is a shrub native to west central Africa. It has green oval leaves and tiny white, yellow, or pink flowers. The plant is used in religious ceremonies to communicate with ancestors. In lower doses, it is used to stave off fatigue and hunger. More recently, anecdotal reports in the West indicate possible anti-addictive effects of iboga.

Active Chemical in Iboga and How It Works

The main hallucinogenic chemical in iboga is called **ibogaine.** Its effects in the brain are numerous and complex, but it does seem to stimulate the serotonin system in the brain similarly to LSD. Ibogaine also blocks a receptor in the brain for the neurotransmitter glutamate, called the NMDA receptor. This effect is unlike that of LSD, but more like PCP or Special K (ketamine). The significance of this is uncertain, but it may relate to ibogaine's reputed anti-addictive effects.

Effects of Iboga

Iboga has potent, long-lasting hallucinogenic effects, typically in the range of twenty-four to thirty-six hours. At low doses, people experience a stimulant effect, while higher doses cause hallucinations, during which the person often stares blankly into space and collapses into a catatonic-like state. During this time, it is believed that the person's soul has left the body and entered the spirit world. Users have reported feelings of levitation or floating, and seeing a rainbow-like aura around objects. Short periods of time may seem extremely long. There is also a sense of detachment, as if users are observing their own actions. Some users report panoramic recall memories of their lives. Traditionally, users communed with their ancestors, which probably had more to do with the mind-set of the users and their religious context than any specific effect of the drug itself.

Anti-addictive effects. The potential anti-addictive effects of ibogaine were accidentally discovered by people taking the drug while also addicted to other drugs. When the long hallucinogenic experience was over, several users reported that they no longer craved the drug to which they were addicted (e.g., heroin). Animal research has supported this possibility, since ibogaine reduces cocaine use by mice, and it suppresses morphine withdrawal symptoms. It may even reduce addiction to tobacco. How these anti-addictive effects occur is not fully known, but there are several possibilities. It may not be a single mechanism, but rather a combination of many effects in the brain.

No controlled human studies of ibogaine for addiction treatment have been done. An open trial has suggested potential anti-addictive effects. A review of treated cases concluded a 10 percent success rate at two years following a single treatment. In 1996, the National Institute on Drug Abuse discontinued clinical trials of ibogaine, so research is currently done by researchers outside of the United States.

The results of ibogaine studies need to be confirmed in controlled trials. The results so far have not been stellar but, more practically, ibogaine may be useful as part of a more comprehensive addiction treatment program. To anticipate a magic bullet for addiction is unrealistic. Addiction has a high rate of relapse over time: features such as social support, counseling, and coping skills are vital for long-term success.

The long-lasting hallucinogenic effects of ibogaine discourage mainstream addiction treatment uses of ibogaine, given the current social and political climate. Also, two deaths of people undergoing ibogaine treatment have been reported. In addition, lab studies indicate possible toxic effects of ibogaine in the brain, if used at high doses. These factors have hindered progress in research for ibogaine. One possible alternative is 18-methoxycoronaridine (18-MC), which is a synthetic drug derived from iboga. It may have anti-addictive effects without causing hallucinations, but much more research needs to be done.

COLLECTIVE ISSUES WITH LSD-LIKE HALLUCINOGENS

Therapeutic Uses

Hallucinogenic drugs were used to facilitate the psychotherapy process, beginning soon after the discovery of LSD in 1949. However, amendments to the Harrison Narcotics Act in 1965 restricted this practice. Analyses of the studies indicated "good/very good" outcome in 40 to 60 percent of cases. Whether this is due to improvement of the therapeutic process or a direct effect of LSD in the brain is uncertain. An uncontrolled study of LSD in depression showed strong antidepressant effects. However, controlled studies have not been done, and arguably cannot be done since a placebo control is impossible. If research subjects don't hallucinate, then they know they got the placebo—in which case it's not a double-blind study.

LSD has been examined for treatment of alcoholism, and may be as effective as other available treatments. Peyote use in the Native American Church is commonly reported to alleviate alcoholism. During the early 1960s, Timothy Leary conducted the Concord Prison Experiment to study the therapeutic use of hallucinogenic drugs in prison inmates. A follow-up study of recidivism suggested there was no long-term treatment benefit, and instead emphasized the value of social support after prison release.

Although formal consensus is lacking, it seems that people do benefit from psychotherapy with hallucinogenic drugs. However, this benefit is not necessarily greater or less than traditional psychotherapy, so hallucinogenic drugs may not offer any clear advantage. Since elaborate safety precautions must be used in such therapy, it becomes apparent why this form of therapy has not flourished.

Hallucinogens and Medications for Mental Illnesses

Not surprisingly, hallucinogenic drugs may interact with medications commonly given for mental illnesses. Tricyclic antidepressants and lithium, used for stabilizing mood in bipolar disorder, enhance the effects of LSD. However, MAO-inhibitor antidepressants and selective serotonin reuptake inhibitors (e.g., Prozac, Zoloft, Paxil, Celexa) decrease the effects of LSD. In some cases, taking selective serotonin reuptake inhibitors completely nullified the effect of LSD.

Do Hallucinogens Cause Insanity?

A few cases have been reported in which excessive use of LSD was thought to cause a lasting psychosis (commonly called "schizing out"). Closer examination of these cases often reveals use of many other drugs besides LSD. Those who did have lasting symptoms had other risk factors and predispositions toward mental illness. It must also be remembered that people who already have schizophrenia sometimes abuse LSD, and some may have schizophrenia that has not yet developed since it commonly emerges in the late teens and early twenties. The rate of people who develop lasting psychosis from taking LSD is the same as the rate of people who develop it in the general population. Not surprisingly, hallucinogens can worsen the symptoms of schizophrenia. A psychological study of long-term religious yage users showed that the users were high functioning and showed no signs of increased mental illness or cognitive problems.

Cognitive Effects of LSD-Like Hallucinogens

It is difficult to measure the effects of LSD on mental abilities. It would not be ethical to give the drug to volunteers over a long term to see if it causes mental impairments. Instead, we can only look at people who have already taken it for a long time, but as a group they may not be the same as the average person. For example, people who abuse multiple drugs have been shown to have anatomical differences in their frontal lobes. In long-term hallucinogen users, mild differences were noted on a test of reasoning, but their scores were still well within the normal range. Further, scores were not different on numerous other measures of mental function.

THE NIGHTSHADE HALLUCINOGENS

Several plants in the nightshade (or Solanaceae) family have hallucinogenic effects. Namely, these are **belladonna** *(Atropa belladonna),* **henbane** *(Hyoscyamus niger),* **mandrake** *(Mandragora officinarum),* jimsonweed or **thorn apple** *(Datura stramonium),* and **scopolia** *(Scopolia carniolica).* Several close relatives in the nightshade family found in South America include **angel's trumpet** (*Brugmansia* species).

These plants differ greatly from the LSD-like hallucinogens. They create hallucinations, but of a more delirious and sedating quality. They also have a different profile of side effects, which are discussed later. The nightshade hallucinogens also have chemical effects in the brain that make a person's memory for the hallucinations poor.

Belladonna

Belladonna is a perennial herb with a straight woody stem, oval, pointy leaves, and solitary yellow flowers. It is native to Europe and northern Africa, but now grows in the United States. The name *belladonna* comes from the Italian words meaning "beautiful woman." Extracts of the plant were historically used to dilate women's pupils, thought to enhance attractiveness by making them seem more excited and interested. The name *atropos* come from the Greek Fate who cuts the thread of life, referring to the plant's potential deadly effects.

Henbane

Henbane is an herb growing wild in Europe, western Asia, and northern Africa. It was known to the ancient Egyptians and mentioned by the ancient Greek physician Pedanius Dioscorides. It has been suggested that the Oracle of Delphi in ancient Greece inhaled smoke from henbane seeds to induce a prophetic trance state. The plant is poisonous to livestock animals, as indicated by its common name, henbane. Its botanical name, *Hyoscyamus,* means "hog bean."

Mandrake

Mandrake was also known to the ancient Egyptians and has been found in burial tombs. The name refers to the humanlike form of the root, which led early Europeans to believe that it could cure human ailments. Mandrake was believed to grow at the foot of gallows, emerging from the semen of men who were hanged. When mandrake was uprooted, evil spirits were believed to make a shrieking sound.

Thorn Apple

Jimsonweed or thorn apple is probably the best known of the many datura species. Datura is native to many parts of the world, including Europe, North America, northern Africa, and eastern and southwestern Asia. The word *datura* comes from its Sanskrit name, *dhatturah. Stramonium* comes the French word *stramoine,* meaning "stink weed," referring to the unpleasant smell of the plant. It was grown in England by seeds imported from Constantinople and was known to settlers in Virginia. Jimsonweed is a variation of the name Jamestown weed, referring to the accidental poisoning of British soldiers in 1676 stationed in Jamestown, Virginia.

Angel's Trumpet

Several members of the nightshade family (*Brugmansia* species) grow in tree form in South America. They are called angel's trumpet because of their large, hornlike flowers. Indians of the Sibundoy Valley in Colombia use these plants for spiritual and religious purposes. They are also commonly kept in the United States as ornamental plants.

Historical Uses: Witchcraft, Lycanthrophy, and Shamanism

The nightshade plants in Europe and North America historically were commonly associated with witchcraft. Witch's brew is believed by some historians to be a salve made with nightshade plants by cooking them in animal fat, which was spread on the skin. The best locations for these salves to absorb were the armpits and vagina. The popular notion of a witch riding a broomstick came from women using a broomstick to apply the hallucinogenic salve to their genitals. The nightshades alter bodily sensations and create a feeling of flying, leading to the legend of witches flying on broomsticks. In reality, they never left the ground, but instead felt and hallucinated that they were flying across the countryside.

Much historical evidence also exists that these drugs were used in sexual orgies in association with Black Masses in medieval Europe. It is hard to know what is truth and what is fiction in this case: some of these reports were "confessed" to members of the Inquisition, who coerced confessions and were notorious for their own preoccupations with sexual matters. On the other hand, taking a hallucinogenic drug by rubbing it on the genitals could certainly encourage sexual hallucinations, if not outright sexual behavior. It is hard for modern people with comparatively relaxed sexual attitudes to imagine what sexual life was like during the Middle Ages. Drugs are often used by people to release themselves from inhibitions conditioned into them by society (e.g., getting drunk), so it is easy to see how drugs such as the nightshades may have served that purpose.

Nightshade hallucinogens were also used for lycanthropy, the belief that people could change from human form into an animal form such as a werewolf. A person would put on a wolf skin and take a nightshade concoction, which would cause hallucinations about changing into a wolf. Descriptions of this practice in ancient Greece very closely resemble the side effects of nightshade hallucinogens: dry mouth, thirst, blurred vision, difficulty swallowing, and staggering walk.

Indians of the Americas used nightshade hallucinogens for more religious reasons. The Aztecs used a plant called *toloache* to prophesy, and it is still used among some Native American tribes. Carlos Castaneda wrote about the shamanic use of datura among the Yaqui in his book *The Teachings of Don Juan: A Yaqui Way of Knowledge.*

Active Chemicals in Nightshade Hallucinogens

The nightshade hallucinogens all contain the same active drug: **scopolamine.** Two closely related drugs, **hyoscyamine** and atropine, are also present, but they only cause bodily side effects and don't enter the brain. All of these drugs block the neurotransmitter acetylcholine in the brain. Acetylcholine is very important for memory, thinking, and other mental functions, so it is easy to see why blocking it causes hallucinations.

Effects of Nightshade Hallucinogens

Nightshade hallucinogens interfere with the autonomic nervous system, altering the automatic bodily functions that it controls. It reduces activity in the digestive tract, and stops production of saliva (causing dry mouth) and sweat. It has mixed and unpredictable effects on the cardiovascular system, leading to changes in blood pressure and heart rate.

Sedation and dreamless sleep are common results of using nightshade hallucinogens. Acetylcholine keeps the brain alert, so when drugs such as scopolamine block it, the result is often sedation. During sleep, acetylcholine also triggers REM sleep and dreaming, so blocking it causes dreamless sleep. Unlike LSD, which activates the brain, the nightshade hallucinogens cause slowed activity, as measured by brain imaging studies.

Both animal and human studies show that nightshade hallucinogens cause short-term mental impairment. As mentioned, acetylcholine is important in many mental functions such as memory. By blocking acetylcholine, the nightshade hallucinogens cause a temporary impairment of memory and other thinking abilities. Accordingly, people usually forget portions of the experiences created by the nightshades. Long-term use of nightshades can cause changes in brain sensitivity to acetylcholine, so mental impairments caused by the drug can take days or weeks to completely disappear.

Abuse and Addiction

Indirectly, the nightshade hallucinogens increase dopamine activity in the brain, so it may have abuse potential since dopamine is released when experiences are interesting or pleasurable. This is another distinction from LSD-like hallucinogens, which do not appear to have any addiction potential. The nightshades have more negative effects on mental functioning and physical health (heart rate, etc.), so continued use of these drugs could be considered abuse. It is ironic that people continue to use them, since they may not even consciously remember much of the experiences.

Side Effects

Side effects of the nightshades result mostly from their autonomic effects, altering vital functions such as heart rate. High doses can cause coma and death by cessation of breathing and cardiovascular collapse.

Skin patches of scopolamine are sometimes used for treatment of motion sickness, but sometimes inadvertently cause confusion and hallucinations. Children may be particularly sensitive to these effects. People with existing mental illnesses should also avoid them.

FLY AGARIC

Fly Agaric

Fly agaric *(Amanita muscaria)* is a mushroom that typically grows in forests. It looks like the stereotypical mushroom that you might see in a cartoon with a white stem and reddish cap with white patches. Some varieties have orange or yellow caps. Amanita mushrooms grow widely across North America, Europe, and Asia. Caution is advised, however, since many similar-looking species of these mushrooms are highly toxic. In North America, fly agaric was used by the Ojibwa Indians of Michigan, and it was used by several tribes across Siberia. Persuasive arguments have been made that it is the divine drug referred to as soma in the ancient Hindu religious text, the Rig-Veda. Although less popular now, fly agaric may be the oldest and most widely used hallucinogen.

Active Chemicals in Fly Agaric

Fly agaric contains many psychoactive drugs, so it's hard to know which one or ones are responsible for the hallucinogenic effects. These include chemicals such as ibotenic acid, muscimol, muscazone, and muscarine. Many of these active chemicals are excreted in urine.

Since the drug is scarce and expensive, many drank their urine to prolong their intoxication. In some cases, the chief of the tribe used the drug first, and others drank his urine to obtain the effects. It's no mystery why this drug never caught on at Woodstock.

How Fly Agaric Works

Each drug in fly agaric has many chemical effects in the brain. Some of these chemicals imitate the neurotransmitters GABA, acetylcholine, and glutamate. Why this should have hallucinogenic effects is not entirely clear. These neurotransmitters are used throughout the entire nervous system, so disturbing their function could easily alter thoughts and perceptions.

Fly agaric seems to be in a class by itself in this regard among plant hallucinogens. However, pharmaceutical hallucinogens that work by interfering with glutamate include PCP (angel dust) and ketamine (Special K).

Effects of Fly Agaric

Fly agaric causes motor dysfunction in both animals and humans, with a general inhibition of motor function, twitching, trembling, and convulsive movements of the extremities. Paradoxically, it has been suggested that Viking warriors used fly agaric to induce a frenzied, enraged state for battle, which gave them the name berserkers.

Like many hallucinogenic drugs, fly agaric causes altered perceptions, a dreamy state, euphoria, and possibly relief from pain. Confusion and disorientation are also reported. The experience often concludes with fatigue followed by sleep.

Side Effects of Fly Agaric

Little is formally known about the toxicity of fly agaric. Ibotenic acid is a strong neurotoxin. This has been demonstrated experimentally, but it is not certain whether neurotoxicity occurs in humans eating the mushroom.

Chapter 10

Cannabis

HISTORY

Cannabis is one of the oldest cultivated nonfood plants, and does not seem to exist anymore in a wild form. It has been used not only for psychoactive effects, but also for its fibers. Hemp fibers have been found in China dating from 4000 B.C., and hemp ropes dating from 3000 B.C. were found in Turkistan. However, it is uncertain whether cannabis was used for psychoactive purposes at those places and times. This may seem ironic, but consider that there are psychoactive plants all around us and most of us are unaware of them.

Although the psychoactive use of cannabis may be a relatively recent practice, it is also ancient. The earliest recorded psychoactive use of cannabis dates from 2000 B.C. in a Chinese pharmacopoeia attributed to a mythical emperor named Shen Nung. He recommended it for absentmindedness and pain, as well as malaria, beriberi, constipation, and female disorders. Hoa-Glio, an ancient Chinese herbalist, recommended a combination of hemp resin and wine as an anesthetic for surgery. But the psychoactive use of cannabis seems to have declined in China by the time of European contact. The Assyrians used cannabis incense as early as 900 B.C. The ancient Hindu Vedas cite cannabis as a divine nectar: the favorite drink of the god Indra was made from it. Indian medical systems cited widespread uses for cannabis, including mania, leprosy, and dandruff. But they also recognized appetite-stimulant and digestive effects. *Bhang, charas,* and *ganja* are all Hindi terms for preparations of the cannabis plant.

The Scythians are credited with spreading the use of cannabis to the Greeks from the steppes. The ancient Greeks knew of the psychoactive effects of cannabis, although it is uncertain how commonly it was used. The Greek physician Galen wrote that cakes made with hemp had intoxicating effects. Women of Thebes were said to use cannabis to dispel sorrow and bad humor. In medieval Europe, herbalists used cannabis for medicinal but not psychoactive purposes. Cannabis was widely cultivated in Europe and later in the American colonies. Hemp fiber was a valuable commodity, used to make rope and canvas sails for ships. Although it is true that many farmers at the time grew hemp, including George Washington, nothing indicates that they were aware of its psychoactive effects.

Cannabis was officially listed in the *United States Pharmacopoeia* until 1937, where it was recommended for a number of therapeutic uses, including sedation. The Marihuana Tax Act, passed in 1937, made it effectively illegal to grow, sell, buy, or distribute cannabis. One of the earliest American studies on cannabis was done on American military personnel who discovered it while stationed at the Panama Canal in the 1930s. From the early 1920s, cannabis gained a public reputation as a menacing deviant drug, an attitude characterized in the movie *Reefer Madness.* By the 1950s and 1960s, cannabis use had gained popularity, particularly among American youths. Use peaked in the late 1970s, when 60 percent of high school seniors surveyed reported having used cannabis, and 11 percent reported daily use. By 1996, 35 percent of high school seniors had used cannabis at least once in the past year, but only 4.6 percent used it on a daily basis. In contrast, use of alcohol and cocaine did not change across this period. In 1990, it was estimated that 6 to 10 million Americans smoked cannabis at least once per week.

The legal status of cannabis in the United States has varied. The Controlled Substances Act of 1970 classified cannabis as a Schedule I drug, which implies that the drug has no currently accepted medical use and has a high potential for abuse. However, a degree of tolerance grew toward cannabis during the 1970s. By 1977, most states classified possession as a misdemeanor, and by 1980, eleven states had decriminalized possession. This atmosphere changed in the 1980s, when severe legal penalties were enacted and the toxic pesticide paraquat was sprayed on domestic cannabis crops.

The Dutch effectively legalized cannabis in 1976. Legalization itself did not increase cannabis usage among the Dutch population, but the advent of commercial access to the drug has. However, parallel increases in use occurred in the United States during this period despite prohibition of cannabis use. Dutch cannabis use has remained equal to or below that of the United States, per capita. How the Dutch example relates to the American legalization issue is still a matter of debate.

Cannabis is a general term that corresponds to three closely related species of plants: *Cannabis sativa, C. ruderalis,* and *C. indica* (see Figure 10.1). *Cannabis sativa* is a tall plant, reaching 8 to 12 feet in height. In addition to psychoactive effects, this is the species that has been a source of fiber for rope and canvas. *Cannabis indica* is found in the Middle East, India, and Central Asia, and is a shorter plant, growing 3 to 6 feet tall. This species has been the traditional source of hashish. *Cannabis ruderalis* is a third species found in Eastern Europe and Central Asia. However, it has little fiber or psychoactive effects.

CANNABIS AND THE HUMAN BODY

Active Chemicals

Cannabis contains approximately 400 chemicals. Some of these are unique to the cannabis plant, and are referred to collectively as

FIGURE 10.1. Two species of cannabis *(Cannabis sativa* and *C. indica).* Copyright 2004 by Marcello Spinella.

cannabinoids. The most potent and most abundant psychoactive cannabinoid is delta-9-tetrahydrocannabinol **(THC),** but there are others (e.g., delta-8-THC, 11-hydroxy-delta-8-THC, etc.). However, not all cannabinoids are psychoactive. For example, cannabinol and cannabidiol are found in the cannabis plant but do not seem to have any psychological effects.

The various names people commonly use to refer to cannabis actually refer to preparations made from different parts of the plant (see Table 10.1). Hashish and charas are a dried resin exuded by the female flowers. (Cannabis plants are dioecious, meaning there are male and female sexes.) This resin has the highest concentration (10 to 20 percent) of THC, so it has the most potent effects. Ganja and sinsemilla refer to the flower buds of the female plants, which averages 5 to 8 percent THC. Marijuana and bhang are from the rest of the plant and have the lowest THC concentration (2 to 5 percent).

Cannabis in the Brain

Cannabis provides one of the more recent and eloquent examples of how studying psychoactive drugs teaches us about the brain itself. As the workings of many psychoactive drugs began to be unraveled in the late twentieth century, some researchers turned their attention toward cannabis. Many psychoactive drugs, such as amphetamine or LSD, were found to work through known systems in the brain. They interfered with neurotransmitters with which we were already familiar. However, the mode of action of cannabis remained mysterious.

To further add to the mystery, THC was found to have its own unique receptor in the brain. The receptor was isolated and analyzed, but it did not belong to any of the known neurotransmitters. This presented a bit of a puzzle: It would seem highly unlikely that a receptor in the brain existed solely for the benefit of cannabis. Rather, it would

TABLE 10.1. Cannabis preparations and THC content.

Name	Preparation	THC content
Hashish, charas	Dried resin exuded by female flowers	10 to 20 percent
Ganja, sinsemilla	Dried tops of female plants	5 to 8 percent
Marijuana, bhang	Remainder of the plant	2 to 5 percent

be more logical if there were a yet-undiscovered neurotransmitter for the so-called cannabinoid receptor. But for years, no such neurotransmitter was discovered.

Perseverance paid off. In 1992, William Devane and his colleagues discovered a neurotransmitter for this receptor. It was named **anandamide,** which derives from the Sanskrit word *ananda,* meaning "bliss." When anandamide was injected into experimental animals, it seemed to have effects similar or identical to THC. They had, in effect, discovered that the brain has its own version of THC. So, when a person smokes cannabis, the THC imitates anandamide, stimulating cannabinoid receptors in the brain and body. As remarkable as it may sound, this wasn't entirely unheard of. For example, endorphin and enkephalin are really the brain's own morphine, and acetylcholine is like the brain's own nicotine. Much of what we know about these neurotransmitters comes from research that was intended to figure out how the drugs work.

A few other neurotransmitters similar to anandamide were discovered afterward, and collectively they are called the **endocannabinoids,** meaning internal cannabis-like chemicals. One of these, **2-AG** (short for 2-arachidonylglycerol) is 170 times more concentrated in the brain than anandamide, and therefore probably plays a more prominent role. Even more interesting is that these endocannabinoids have been found to play a normal role in memory, thinking, movement, and pain perception. This too is logical, since using cannabis affects all of these functions.

Two different cannabinoid receptors have been found, abbreviated CB_1 and CB_2. Both types are found in many parts of the body, but only the CB_1 receptor is found in the brain and thus responsible for the psychoactive effects of THC. The location of the CB_1 receptors in the brain reveals something about the effects of THC. For example, many receptors are found in the basal ganglia, cerebellum, and cortex, which is where THC apparently has its effects on movement and thinking, since that's what those parts of the brain do. Receptors are also found in the limbic system, which is where THC has its emotional effects. In contrast, very few receptors are found in the brainstem and spinal cord. Since the brainstem controls vital functions, such as breathing, cannabis has little effect on them. Alcohol or heroin, in contrast, will kill a person if too much is taken since they act on receptors in the brainstem.

Drug Issues with Cannabis

Cannabis is most commonly consumed by smoking in a cigarette or pipe. An average cannabis cigarette has about 0.5 to 1 g of cannabis. Considering that cannabis is usually 4 to 5 percent THC, then the average cannabis cigarette contains about 50 mg of THC. However, the amount that actually reaches the blood after smoking one cannabis cigarette varies from 0.4 to 10 mg. Interestingly, people may unknowingly regulate their smoking based on the potency of the cannabis: the weaker it is, the longer and deeper they seem to inhale. Conversely, high-potency cannabis causes people to inhale more briefly and shallowly. The common argument that modern cannabis is more addictive because it has more THC may then be irrelevant, since people seem to adjust their smoking (increase or decrease) according to its potency.

After inhaling the first puff, the psychological effects start within six to twelve minutes. They reach maximum intensity at fifteen to thirty minutes and typically last two to four hours. The lungs and liver break down THC into 11-OH-THC (pronounced "eleven hydroxy THC"), which itself has psychoactive effects. However, this chemical is further broken down and eliminated from the body. The removal process is slow, however, since half of the drug is eliminated every three to five days, and small amounts of the drug or its breakdown products remain in fatty parts of the body for weeks. These remainders are detectable by drug tests, but the amounts are too small to have psychoactive effects. Only in heavy, chronic users could they possibly accumulate enough to have an appreciable effect.

The way cannabis is taken greatly affects its intensity. Cannabis is absorbed into the blood thoroughly when smoked, but poorly when swallowed. For this reason, cannabis leaves are hardly ever eaten by themselves. However, some people have discovered the practice of cooking cannabis in food, such as brownies. This is typically done in foods that are high in fat. Since THC absorbs well into fat, eating it this way allows it to absorb better through the digestive tract.

The Effects of Cannabis

Cannabis stands in a class by itself, not only because of its unique effects on brain chemistry, but also for its constellation of effects. It shares common effects with other drugs such as hallucinogens or sedatives, but its total pattern of effects is unique.

Psychological Effects

As with many other drugs, the psychological effects of cannabis can vary. They depend on how much experience the person has with the drug (if any), what expectations the user has, and even some personality characteristics. In general, using cannabis produces relaxation, a mild lifting of mood, and giddiness (see Box 10.1). The senses are heightened so that smells, tastes, and so on may seem more intense. Time also seems to pass more slowly. However, half of cannabis users report experiencing unpleasant side effects such as anxiety or panic reactions at some time. These reactions are more common in inexperienced users and those using high doses. Other unpleasant effects include dissociative symptoms such as derealization (the feel-

BOX 10.1. The Effects of Cannabis

Subjective effects
- Relaxation
- Mild euphoria
- Giddiness
- Heightened sensory perceptions
- Stimulates appetite
- Perceived slower passage of time
- Panic or anxiety
- Dissociative symptoms

Physiological effects
- Increases heart rate
- Reduces body temperature
- Slowed gastrointestinal function

Brain activity
- Mixed excitation and inhibition
- Increases deep, slow-wave sleep; reduces REM sleep

ing that one's experiences are not real, or are a dream), or depersonalization (feeling out of one's body, or detached from oneself). Reassuring and comforting the person who is having negative reactions seems to be an effective way to treat the problem when it occurs, although anti-anxiety medications are sometimes given.

Physiological Effects

Cannabis causes an increase in heart rate and a reduction in body temperature. It also mildly slows digestion. Cannabis use does not appear to induce a hangover in the way that alcohol or other sedatives can. The hangover after alcohol use has documented effects on sleep, mood, motor, and thinking abilities.

Cannabis interacts with several hormones. It tends to decrease sex hormones such as testosterone, but it's not clear that this amounts to anything significant in normal circumstances. It does reduce sexual function in animals, but whether it has an effect in humans at the doses normally used has not been studied. Cannabis also causes a release of stress hormones. Many psychoactive drugs are known to have a similar hormonal effect, including ginseng and caffeine.

Cannabis is widely known to stimulate appetite, what's known as "the munchies." This occurs in both animals and humans alike. Most likely, it happens through THC's effects in the hypothalamus, which regulates appetite. This effect of cannabis has been considered as a therapeutic use for the drug for patients with poor appetite.

Brain Activity

Cannabis has been shown to cause changes in brain activity, whether measured through the brain's electrical activity, blood flow, or metabolism. Monkeys given THC show a slowing of electrical activity in the brain for a few hours (accompanied by sedation), followed by a period of increased activity, suggesting a rebound excitability when the drug has partly worn off. Long-term use of cannabis in humans is associated with slowed electrical patterns across the frontal lobes of the brain.

Studies of the effects of THC on blood flow and metabolism in the brain in humans produce very mixed results, sometimes increasing brain activity and sometimes decreasing it. This may seem confusing, but it makes sense when you consider that the psychological effects of cannabis are also very variable, depending largely on one's mind-set and environmental setting.

One interesting finding is that THC alters activity in the cerebellum. The cerebellum is known to play a role in the experience of time, and an altered sense of the passage of time occurs when people use cannabis.

Sleep

Cannabis has effects on sleep, shown by electrical patterns in the brain (measured with the electroencephalogram). When people are given daily doses of THC, they spend less time in REM sleep and more in Stage 4 sleep. REM sleep is an active, dreaming state and Stage 4 is a deep, restful state, so it seems that THC increases restorative sleep. However, the doses used in this study were large (70 to 210 mg) compared to the dose received in a usual cannabis cigarette (about 10 mg). So we don't know if the effects caused by more typical use would be as extreme.

Movement

Studies of THC in animals and humans show similar effects on the motor system. THC makes it harder to maintain balance and reduces fine motor control and dexterity. Some report an increase in reaction time. Extremely high doses given to animals cause immobility and abnormal posture.

A primary concern is whether cannabis affects the ability to drive, which requires good motor functioning. Surprisingly, cannabis seems to have relatively small effects on driving ability. Nonetheless, any impairment while driving is potentially dangerous, considering the consequences. Thus, driving is best avoided when cannabis is used, just as it is best avoided when taking a sedating allergy medication. Cannabis certainly should not be combined with alcohol by anyone who is going to drive, since the two may magnify each other's effects.

Cognitive Functions

Cannabis certainly affects higher thinking abilities. However, the results of this type of research must be handled appropriately to avoid false conclusions. When looking at a cannabis study, one must first consider the subjects who are being studied (e.g., occasional cannabis users, frequent users, or first-time users). Do these subjects have any preexisting problems that could affect the results? For example, is it a healthy group of responsible adults, or do the subjects have any psychiatric or criminal history? Also important to consider is the amount of cannabis (or pure THC) that is given to them, the duration of time it is given (e.g., is it in the range of what is normally used), and what is used to measure the effects (e.g., what tests or behaviors are being measured). Many studies of the same drug produce different results because they use very different methods to study it.

It is also very important to distinguish between *correlational* and *experimental* studies that examine cannabis. In an experimental study, researchers compare the people who get the drug to people who get the placebo. That way they know that the difference is due to the drug and not to expectations. In correlational research, a relationship is shown between two variables. For example, it has been shown many times that marijuana use correlates with juvenile delinquency. But a correlational study only shows a relationship, and cannot prove which causes which, even though such statistics are often misused for such purposes. So juvenile delinquents are more likely to smoke marijuana than others, but nothing suggests that smoking marijuana caused the juvenile delinquency. There are dozens of other equally probable possibilities. For example, juvenile delinquents tend to disregard rules and may just be more likely to break laws and use illegal drugs in general. Correlational research is quite often misrepresented in the news, and sometimes even by scientists who should know better.

Several studies show that cannabis decreases attention and concentration. This partly depends on the dose of THC. In one study, impairment of attention occurred with a higher dose of THC, but not with a lower dose. The most basic forms of attention do not seem to be impaired, but impairment can occur for more complex forms of attention, such as shifting attention. Cannabis has a more noted effect on working memory. Working memory is a form of short-term memory, enabling a person to hold information in mind and work with it (e.g., mentally adding up prices while you are shopping). So, for example, cannabis causes a person to remember fewer words on a short-term memory test.

The effects of cannabis on long-term memory have been examined in numerous studies, again with variable results. Cannabis impairs some forms of long-term memory, such as the ability to recall words from a list. However, it does not affect the ability to learn pairs of words. Cannabis may partly disrupt the brain's process of making memories permanent. In contrast, cannabis does not affect the ability to recall information already permanently stored.

Cannabis use is sometimes cited as the cause of an amotivational syndrome, consisting of a loss of interest in social activities, school, work, or other goal-directed activities. However, no evidence supports any causal relationship. Instead, evidence suggests that the amotivational syndrome symptoms are part of depression. In contrast to alcohol, no evidence supports the idea that cannabis causes an increase in violent behavior. However, cannabis use is not a good idea for people with existing psychiatric disorders such as bipolar disorder or schizophrenia.

Daily cannabis users show more mental impairments than infrequent users. However, these studies are problematic since the subjects are self-selected, so we cannot determine whether cannabis causes the impairments. It may well be that people with mental impairments are more prone to heavier cannabis use. Indeed, people who abuse multiple drugs have been shown to have smaller frontal lobes than nonusers (Liu et al., 1998). Although long-term users tend to experience greater difficulty in complex forms of memory and attention, these deficits are relatively subtle.

The effects of THC on mental abilities such as attention are comparable to the effects of alcohol. Both THC and ethanol increase reaction time. The effects of THC tend to last longer than those of alcohol. Of course, using THC and alcohol together produces additive effects.

In conclusion, there is good evidence that cannabis produces impairments in attention and concentration, the formation of new memories, perceptual-motor function, and executive functions. However, these impairments are relatively modest, comparable to those of alcohol. As with alcohol, the degree of impairment depends on how much is used.

CANNABIS ADDICTION

Chronic use of cannabis causes cannabinoid receptors to adapt and become less sensitive (tolerance). This effect is not permanent, and reverses after discontinuation. As with any mood-altering drug, cannabis has the potential to produce dependence. It is likely that dopamine is involved in the reinforcing effects of cannabis. However, cannabis is arguably less addictive and produces milder withdrawal symptoms than drugs such as cocaine, alcohol, and heroin. The intensity of the withdrawal syndrome would be related to the degree of use and tolerance. There is evidence that chronic THC administration can increase ethanol intake.

Suddenly stopping chronic cannabis use can produce a withdrawal syndrome consisting of restlessness, irritability, insomnia, nausea, and muscle cramping. However, this syndrome occurs only in people who use high daily amounts and suddenly stop. These full-blown symptoms are not usually seen in clinical populations, and frequent users of cannabis are not driven by a fear of withdrawal as opiate addicts may be. **Dronabinol** (Marinol) is a synthetic oral form of THC developed for medical use. A thorough study concluded that dronabinol has very low abuse potential since people tended to use it in the recommended doses for the recommended period of time, and there was no evidence of "script chasing" or "doctor shopping."

THERAPEUTIC USES OF CANNABIS

Cannabis shows promise for clinical treatment of several conditions, including nausea, pain, spasticity, multiple sclerosis, glaucoma, and AIDS wasting syndrome. The legal status of cannabis is currently a gray area and a matter of constant legal debate.

THC was first tested for the reduction of nausea and vomiting during chemotherapy in the 1970s (Vincent et al., 1983). THC is superior to placebo and as effective as or superior to the pharmaceutical antinausea drug prochlorperazine (Compazine). People taking THC reported more side effects, but these effects did not reduce patients' preference for the drug. The intensity of the antiemetic effectiveness of THC correlates with the subjective high reported.

Appetite Stimulant

Some evidence suggests that THC is effective in reducing nausea associated with cancer chemotherapy and stimulates appetite. In patients with appetite loss due to advanced cancer, THC (2.5 mg taken orally, three times per day) stimulated appetite and was well tolerated.

Cannabis and cannabis-derived drugs also are effective for AIDS-related appetite loss and wasting. Open trials were conducted with positive results, and were followed by controlled trials. The synthetic THC drug, dronabinol (2.5 mg taken orally, twice daily), improved appetite in patients with AIDS. Mood also improved, and stable body weight was maintained for at least seven months. Increases were also seen in body fat (1 percent). The psychological effects of dronabinol can reportedly be curbed by using another drug called prochlorperazine.

Glaucoma

Glaucoma is a condition involving excessive fluid pressure in the eye, eventually impairing vision. A controlled study in the early 1970s showed cannabis to reduce eye fluid pressure in people suffering from glaucoma. THC decreased pressure by an average of 37 percent, with maximum decreases of 51 percent. How THC achieved this effect is uncertain.

Movement Problems

In studies, THC reduces muscle spasticity in both animals and humans. A 10 mg dose of THC significantly reduces spasticity resulting from many different illnesses. A survey of multiple sclerosis patients indicated that cannabis relieves spasticity, pain, and tremors caused by the disease. A study in mice similarly supports a role for THC in reducing the tremor and spasticity caused by multiple sclerosis. A recent study in people with Parkinson's disease showed that THC may reduce the excessive movements caused by the medications used to treat the disease.

Side Effects

Pregnancy

Cannabinoids are teratogenic and should be avoided during pregnancy. When women use cannabis during pregnancy, it's associated with a risk of premature delivery and low birth weight. Babies of mothers who smoke cannabis during pregnancy can experience mild withdrawal symptoms after birth. Problems with language, attention, memory, and behavior are related to prenatal exposure to cannabis.

Pulmonary Toxicity

One of the major health problems associated with cannabis use relates to the fact that it is most often smoked. Cannabis and tobacco smoke, apart from containing different psychoactive drugs, are actually very similar in their composition. Cannabis smoke is carcinogenic. It is at least as harmful as tobacco smoke, and contains three times as much tar and five times as much carbon monoxide. Cannabis smoke irritates airways, and reduces the capacity for breathing.

Cardiovascular System

THC commonly increases heart rate and blood pressure. These effects are not generally severe. Cannabis has a high safety margin, and no fatal dosages have ever been reported in humans. However, these effects could be a danger to someone who already has high blood pressure or heart disease. THC dilates blood vessels in the eye, creating the red eye commonly associated with cannabis use.

Immune System

The immune system has cannabinoid receptors, which means that THC could theoretically suppress the immune response. However, the relevance of these effects appears to be subtle or insignificant.

Brain Toxicity

There is a common conception that cannabis use kills brain cells, but this is actually a controversial topic. Large doses over a long period of time kill neurons in animals, but most drugs given in unrealistically high amounts will do the same. What is more relevant is what happens at the doses of THC normally used by humans.

Until recently, little clear evidence showed cannabis use is toxic to neurons. However, a recent study showed that THC damages neurons in the hippocampus of primates in doses that are proportionate to normal human doses. This is likely due to certain chemical changes and free radical damage. The toxicity of THC was prevented by giving nonsteroidal anti-inflammatory drugs (such as aspirin) and the antioxidant vitamin E. However, THC is not the only chemical in cannabis. Cannabidiol, for example, actually has protective effects on neurons. So the total combined effect of using cannabis is still uncertain. Limiting use of cannabis and using antioxidants may be the best strategy to avoid potential harm.

Seizures and Epilepsy

Cannabis may have harmful effects in people with epilepsy. Drugs in cannabis have complex and mixed effects on seizures, and can sometimes trigger them. Cannabis tends to slow activity in the brain, which is generally good for seizures, but when the drug wears off, the brain can rebound and become overexcitable. This effect of cannabis can last twenty-four hours or longer after cannabis is taken. It would depend also on how much was used and over what period of time. The effects of cannabis on seizures are, at best, unpredictable, so people with seizure disorders would probably do best to avoid it.

SUMMARY

Cannabis has been used by humans for thousands of years for both psychoactive and nonpsychoactive purposes. It remains in popular use for recreational purposes, and potentially for medical purposes. Through studying cannabis, we have discovered new chemical systems in the brain and body, and we are just beginning to appreciate how they work.

To be certain, cannabis is not a harmless drug. Like any psychoactive drug, it should be treated with respect and caution. It has prominent effects on thinking, emotion, the senses, and coordination. Reckless use of any psychoactive drug is a recipe for disaster.

However, the case for cannabis impairing mental functions has been overstated in many studies. Cannabis certainly impairs mental abilities, as do other drugs such as alcohol, codeine, and antihistamines. The degree of impairment is determined by how long and how often a person uses the drug. As with alcohol, cannabis should certainly not be used by adolescents or pregnant women due to potentially harmful effects on development. Also similarly to alcohol, cannabis use is not appropriate when peak mental abilities are needed, such as when driving.

Cannabis carries some potential for dependence and addiction. Compared to cocaine, heroin, alcohol, and nicotine, cannabis may have less addictive potential and withdrawal effects. Nonetheless, some people develop compulsive and maladaptive patterns of use that are harmful and require treatment. Individuals with existing mental disorders (such as schizophrenia or bipolar disorder) should be leery of using cannabis in the interests of avoiding compulsive use patterns.

Cannabis has many potential medical uses, ranging from treatment of cancer to AIDS, multiple sclerosis, muscle spasticity, pain, and glaucoma. More controlled research is warranted in order for cannabis to become a legitimate treatment for these conditions. Cannabis is currently classified as a Schedule I drug by the DEA, suggesting that it has no legitimate medical uses. With the proper research, it is reasonable to suggest that it could be switched to Schedule II, where it could be prescribed and its use monitored by a physician. Cannabis may produce some mental impairments and have abuse potential, but so do opioid drugs such as codeine and Darvocet, yet we tolerate these drawbacks when they are used properly. Thus, cannabis could be used in a controlled manner similar to opioids.

Whether cannabis should be legalized for recreational use is a separate matter. One could argue that alcohol is a legal drug that also produces mental impairment. Both alcohol and tobacco have known addiction potential, yet they remain legal. Whether another such drug should be added to the list of legal drugs is a matter of philosophical standpoint. Ultimately, whether the use of cannabis is to be permitted and for which purposes (medical or recreational) remains to be collectively decided by society and the legislators we appoint. Hopefully, these decisions will be made on the basis of opinions formed by sound scientific research.

Chapter 11

Use, Abuse, and Addiction

Drug addiction is an affliction that has affected countless millions of lives. It is both stubborn and persistent, and despite progress in the fields of psychology and neuroscience, we are no closer to eradicating it today than we were a hundred years ago. It is by no means a new problem, but probably has existed to some degree as long as humans have had access to psychoactive drugs.

It seems that everybody has either struggled with addiction or is close to someone who struggled, or continues to struggle, with it. The National Institute on Drug Abuse confirms this picture. In 1999, 14.8 million Americans had used an illegal drug at least once in the month before being interviewed. About 3.5 million were dependent on illegal drugs and another 8.2 million were dependent on alcohol. Given a population of about 270 million in the United States in that year, it means that roughly 4 percent of the population was addicted to a drug. The trends for different drugs and geographic areas tend to fluctuate: several cities in the northeastern United States saw a decrease in the use of cocaine and crack in 1999, but heroin use had increased among eighteen- to twenty-five-year-olds by 51.4 percent from 1997 to 1999. However, the use of both cocaine and heroin have increased since 1999, as has the number of emergency room reports of treating patients who had used these drugs. Marijuana use seemed to have stabilized for the time being, while use of Ecstasy (MDMA) had been growing in recent years. Overall, estimates from 2002 indicate that 8.3 percent of Americans used an illicit drug in the month prior to the survey. This percentage was highest (20.2 percent) among the eighteen- to twenty-five-year-old group.

These statistics are disturbing. They emphasize the point that drug use is widespread, with no sign of slowing down. Some successfully quit, while others relapse, and new users continue to appear. In many respects it is a frustrating problem: it seems as if there is no cultural memory from which mistakes of the past can be learned and avoided. Instead, it seems as if newer generations continue to suffer the same problems as those before them.

Why is addiction such a stubborn problem? Why have our best efforts failed to make a serious dent in it? This chapter takes a look at what addiction is, why it's so persistent, and what can be done about it.

DEFINITIONS

Before trying to explain addiction, it's important to have a firm idea of what it is. We must understand the distinction between drug use, abuse, and addiction. In reality, the lines between them are fuzzy, but definitions help us understand and agree on what it is.

Good versus Evil

Who gets to decide whether taking a drug is use, abuse, or addiction? Certainly not all drug use is abuse. So who has that authority? The standard manual for defining and diagnosing mental illnesses is the *Diagnostic and Statistical Manual of Mental Disorders,* Fourth Edition, Text Revision (DSM-IV-TR), published by the American Psychiatric Association (2000). They list criteria for drug abuse and addiction, which are discussed in this chapter. The symptoms chosen to define addiction are very common themes in drug abuse and addiction, and most will agree that they paint a fairly accurate picture of the condition. But rather than just appealing to authority, it's important to know why they chose the symptoms they did. Beneath all of these criteria is an underlying principle that we use to decide if use of a drug is "good" or "bad."

The underlying idea is to weigh costs and risks against benefits. We must first consider the total costs and risks of taking a drug in terms of how it affects one's financial, physical, mental, and social well-being. This is then weighed against the potential benefits of the drug, or how it might improve a person's well-being. What tips the scale from use to abuse is the relative relationship of costs and risks versus benefits. In a larger sense, everything that we do involves some risks and costs. Most of what we do involves some form of benefit. Whether any given activity is worthwhile, then, is a matter of how much good it creates relative to how much harm (or potential for harm). Whether or not we realize it, we make these types of judgments all of the time.

It's easy to see, in this framework, why taking a painkiller after surgery is acceptable. Pain after surgery can be severe, which affects a patient's mental and physical health. The limited use of a prescribed painkiller presents minimal risks, reduces suffering, and greatly enhances the person's ability to function. Medical examples such as this are fairly clear-cut, or so it would seem. Now consider the example of a heroin user injecting heroin intravenously. Chronic heroin use causes physical and mental health problems, and stopping it causes a serious withdrawal syndrome, making the user dependent on it. It's exceedingly expensive; frequent use is a financial drain as well. Clearly, recreational use of heroin is a case where the costs and risks far outweigh any benefit.

The tricky part involves the cases somewhere in between. What about a person who comes home from work and drinks one or two glasses of wine? Assuming that the person isn't driving and is able to carry out normal duties at home, is it any harm? Research says that if a person has one or two alcoholic drinks per day (on average, and depending on one's body size), the odds for being healthy are actually improved. Moderate alcohol use is associated with a reduced rate of heart disease, stroke, stress, depression, dementia, ulcer, gallstones, rheumatoid arthritis, osteoporosis, and type II diabetes mellitus. But it's a fine line. Any more than one or two drinks and the risks overshadow the potential benefits.

Use versus Abuse

Use of a drug simply means taking it to intentionally produce a desired effect, whether for therapeutic or recreational purposes. This includes taking aspirin to relieve a headache, taking insulin to treat diabetes, or drinking alcohol at a cocktail party. The term *use* doesn't make any judgments, either positive or negative.

The phrase **substance abuse** immediately carries a negative connotation. There are four criteria for what constitutes drug abuse, and the presence of any one of them signifies abuse (see Box 11.1). The first deals with repeated drug use that makes a person unable to fulfill important roles, such as performance at work, school, or in the home. This criterion is not enough by itself, since many people who abuse drugs remain generally functional—although it could be argued that if you look hard enough, you'll find the problems. The second criterion states that abuse occurs when a person uses a drug in situations in which it's dangerous. Immediately, driving a car while drunk comes to mind, but this applies to other situations as well. Riding a bicycle while drunk can be equally dangerous to the person who's drunk. Third, having legal problems as a result of drug use also signifies abuse, although this becomes a matter of which drugs a society considers legal and illegal. Cannabis use is illegal in the United States, but legal in Holland. Obviously, getting on an airplane and flying to Amsterdam doesn't change much except the legal status of the drug. Fourth, abuse is indicated when drug use continues despite persistent social problems. When drug use starts to affect a person's relationships, leading to disagreements, friction, arguments, or even fighting, it's a problem.

Addiction

At some point, drug abuse crosses the line into drug addiction. Seven criteria describe when drug use is considered to be **substance dependence** (see Box 11.2). A person must show three or more of these criteria for a twelve-month period. Of course, the twelve-month

BOX 11.1.
American Psychiatric Association Criteria for Drug Abuse

1. Recurrent drug use and failure to fill important roles in the person's life
2. Use of a drug in situations where it's dangerous
3. Legal problems resulting from drug use
4. Continued drug use despite persistent problems

BOX 11.2.
American Psychiatric Association Criteria for Substance Dependence

1. Tolerance
2. Withdrawal syndrome
3. Taking larger amounts over time, longer than intended
4. Desire or unsuccessful attempts to cut down or quit
5. Excessive time spent getting, using, or recovering from a drug
6. Normal activities hindered by the drug use
7. Persistent use despite knowledge of harm

rule is just for purposes of classification. Obviously someone who is dependent on a drug for eight months still has concerns.

The first criterion is tolerance for the drug, which is discussed in more detail in Chapter 3. Essentially it means that taking the same dose over time has a progressively smaller effect, so to maintain the same intensity of effect, the dose must be increased. A withdrawal syndrome is the second criterion, meaning that a person cannot function normally without the drug. This may be different for every drug, and typically it consists of the opposite effects of the drug. For example, cocaine causes euphoric happiness and cocaine withdrawal causes depression. Sometimes people take similar drugs to ease withdrawal symptoms. For example, people may take other sedatives when quitting alcohol. The third criterion is that a user ends up taking a drug in larger amounts or longer than intended. Often, people realize they are using a drug too much or too often, but their attempts to quit are unsuccessful, which is criterion 4. Criterion 5 reflects the fact that a person who is addicted often spends considerable time getting the drug, using it, or recovering from its effects. Time spent on the drug is time that is not spent on other activities, which is criterion 6. A person addicted to a drug will sometimes avoid contact with people, including activies such as family events, work, or socializing with friends. The last criterion explicitly addresses the risk-benefit aspect, when the person continues to use the drug despite knowing that it causes mental or physical harm.

EVOLUTIONARY BASIS FOR DRUG ABUSE

At first glance, it would not seem that evolution has much to do with addiction. However, considering human psychology in terms of our origins sheds light on drug use and abuse. Humans essentially have not changed for tens of thousands of years in terms of our physical or psychological makeup. We have more knowledge nowadays, since it accumulates through language and culture. But if you could go back in time and kidnap a human baby from 10,000 years ago and raise it in today's world, the baby would most likely be just like any other.

For most of human history, we've lived as hunter-gatherers, and evolution has selected traits in us that make us successful candidates for living that lifestyle. The food our ancestors ate was whatever animals they could kill or whatever plants they could find, so they needed the physical and mental abilities to get that food. They lived in small nomadic groups or tribes, making social skills a necessity. Aggression was needed at times to fight for a mate or food, or to protect the tribe from others. The only psychoactive drugs our ancestors consumed came from plants. Perhaps occasionally they stumbled upon a plant that had noticeable psychological effects, and sometimes perhaps they remembered the plant and went back to use it.

Even though our bodies and minds are still designed for hunter-gatherers, the situation has changed. Agriculture and language gave rise to civilization, and most humans live a life that is very different now. This is called **evolutionary mismatch:** evolution prepared us for one set of circumstances, but rapid changes in the past few thousand years have led us to live otherwise.

Now consider modern, industrial life in contrast to our ancestors' lives. The food we eat comes from agriculture and food is no longer scarce but in abundance. Food is created for its taste more than its nutritional necessity (think McDonald's), and as a result we have increased illnesses such as obesity, diabetes, and heart disease. Humans were sparse thousands of years ago, but now we live in cities with over 50,000 people per square mile. The consequences of crowding and stress are evident and our tolerance for social contact is taxed to the limit. The same instincts that drove our ancestors to fight off an invading tribe are now at work when we create nuclear weapons. This is not to argue that civilization is inherently evil (although many people have), only different from what evolution prepared us for. For all its shortcomings, it also has its benefits. After all, our ancestors likely lived harsh lives and were lucky to see age thirty.

Now consider the psychoactive drugs available today. The first consideration is variety: a greater variety of drugs is available to us than has ever been available to humans. Our ancestors would have been lucky to stumble across two or three species of psychoactive plants. We, on the other hand, have ready access to hundreds of legal and illegal psychoactive drugs. The second difference is potency. Humans in South America only had coca leaves, but for the past 150 years, we've had pure cocaine. The Inca used coca for thousands of years without apparent detriment, but cocaine has been a liability ever since it was purified from the plant. A third difference in modern drug use is the way we take drugs. Eating a plant or drug has mild effects, since it absorbs slowly into the body, and excessive amounts will cause us to self-protectively vomit it up. However, we've invented smoking and injection, so now the same dose of drug hits the brain faster and creates more intense effects.

So we're taking drugs in ways that evolution never prepared us for. Nothing in human history could have prepared us for the effects of smoking crack. Yet these drugs and methods are available now, and we must deal with them. Again, this is not to say that all drug use is bad. It's only different from what evolution has prepared us for, and we must keep this in mind when we make decisions about what to use and how. When you think about legalization of drugs such as heroin in this context, it seems absurd.

PSYCHOLOGICAL PERSPECTIVE

Drugs and Behavior

Many areas of psychology can shed light on addiction. Perhaps the best place to start is a behavioral perspective. Behavioral psychology classifies influences on behavior as **reinforcement** or punishment, where reinforcement increases a behavior and punishment decreases a behavior. For example, taking a drug may induce pleasant feelings, or it may take away pain or feelings of nervousness. Both of those consequences are reinforcement for taking the drug, making it more likely that the drug-taking behavior will be repeated in the future. Punishment can also encourage drug use when attempts to stop taking a drug have aversive effects. For example, quitting tobacco is difficult for many because of the mental and physical withdrawal symptoms that occur.

On the other hand, the unpleasant consequences of a drug, such as feeling nauseous after drinking too much alcohol, also serve as punishment. This type of punishment makes it less likely that the drug will be taken again in the future. All drugs have both pleasing and aversive effects, so the drugs that are taken repeatedly are those whose pleasing effects are strongest relative to their aversive effects. For example, aspirin takes away pain and causes mild side effects in most people. As a result, aspirin has been very commonly used. The most addictive drugs, such as cocaine and heroin, are so intensely reinforcing that users often overlook their negative effects.

The reinforcing or punishing conditioning that occurs with drugs does not just come from the drug itself. Much of it comes from social influences, where some drugs are encouraged and others are discouraged. Drinking a glass of beer in a social gathering on a weekend may be encouraged, while drinking hard liquor first thing in the morning is usually discouraged. The use or possession of many illegal drugs carries a penalty of jail time, which serves as a strong potential punishment.

However, learning drug use and related behaviors occurs in many instances before reinforcement or punishment ever occur. Many people make choices about using illegal drugs without having ever been sent to jail. In psychology, this is called social learning or observational learning, since it can occur simply by observing and interacting with others. So people who observe a family member abusing drugs while growing up may internalize many of those behaviors long before they ever repeat them (or avoid repeating them).

Addiction and Emotion

In addition to considering behavior, we must consider the effects of drugs on emotion, since emotions are what motivate us to act, or not act, as the situation would have it. Most psychoactive drugs are used to change one's emotional state, either directly or indirectly.

Alcohol can create feelings of relaxation, reduce tension, and lift up one's mood. Caffeine mildly lifts mood as well, although larger amounts will induce anxiety. Cocaine and heroin both induce intense euphoria, despite the fact that cocaine stimulates the mind and heroin sedates it. Although aspirin has little direct effect on mood, it relieves pain, which indirectly reduces emotional suffering and distress.

Thus, many psychoactive drugs are shortcuts for changing our emotional state. Changing our emotional state by changing behavior takes much more work and time to accomplish. Although the results of changing behavior are better in the long run, drugs are alluring because they offer much more immediate results. Consider the example of a person who is feeling sad or depressed. There are many ways he could make himself feel better. Exercising has proven mood-lifting results, equivalent to antidepressant drugs in some studies. Making plans with friends and increasing social contact is beneficial to mood. Getting adequate rest and nutrition are also important. In more severe cases, he could talk to a counselor and learn to handle stressors differently, or reevaluate social ties with friends and family.

These methods are reliable for improving mood, but they require persistent effort over a period of time. Depression, unfortunately, often doesn't depart quickly. On the other hand, an array of psychoactive drugs can also alleviate depression, even if only temporarily. For example, alcohol, cocaine, opiates, and cannabis can lift mood, or at least reduce the feeling of suffering. Unfortunately, the effects of these drugs are not long-lasting. Typically, the depression returns or may worsen when the effects wear off. Depression is only one example. People can similarly use a drug to quell tensions, tranquilize anger, or alleviate stress.

So drug addiction can be viewed as a maladaptive form of mood management, although people are often unaware that they have fallen into this pattern. Reliance on a drug to manage mood makes people less able to manage their moods on their own, by altering behavior and forming healthy habits. Mood management through drugs may be socially tolerated in some situations, such as moderate alcohol consumption at a social gathering or coffee to brighten up a morning. However, addiction is the same pattern to a more harmful degree.

Stress appears to play a significant role in drug abuse. It's a significant factor in the development, continuation, and relapsing of addictions. For example, stress is a major contributor to smoking relapse, even long after a person has quit. Children exposed to severe stress (such as abuse or the loss of a parent) are at increased risk of drug abuse during adulthood. Stress as an adult can contribute to addiction as well. Between 30 and 60 percent of people with an addiction also have post-traumatic stress disorder (PTSD). A startling 75 percent of combat veterans with PTSD suffer from alcoholism.

Addiction and Thinking

There's a strong two-way relationship between thoughts and emotions. Thoughts can trigger emotions, and emotions can shape (or sometimes warp) our thinking. People with mood disorders such as depression and mania not only have emotional symptoms, but changes in their thinking as well that accompany the extreme moods. In mania, a person's judgment is affected so that they tend to be unrealistically optimistic, while in depression people tend to be overly self-critical and pessimistic.

Not surprisingly, psychoactive drugs have effects on people's thinking, often by altering emotional states. One of the core features of addiction is the persistent use of a drug despite negative consequences for one's physical, mental, or social well-being. The musician Elton John stated that he had continued to use cocaine even after experiencing seizures that were caused or worsened by the drug.

The great irony is that people who are addicted often have poor awareness of the problem, although it may be blatantly obvious to those around them. It's not a matter of intelligence: people with addictions don't have a lower IQ than those who have never developed an addiction. But something about the addictive state limits a person's awareness of the problem. In a sense, unawareness of the problem is itself the problem, creating a catch-22. The powerful emotional effects of certain drugs naturally influence users to think in positive terms about the drug use, or to avoid negative thoughts about the drug, despite evidence to the contrary.

Addiction and Personality

Personality traits, which are characteristics that tend to be fairly stable over a person's lifetime, can contribute to addiction. For example, nervousness, moodiness, low self-esteem, anger, resentfulness, and personal instability (e.g., employment, relationships) have been associated with addiction. Keep in mind that although someone who is nervous may be more likely to use a drug to calm down, being nervous does not necessarily mean that a person will use the drug or become addicted. Personality traits are tendencies, not destiny. Thus, no clear "addictive personality" exists.

One trait that seems to have particular importance is impulsivity, or the tendency to act on immediate needs rather than long-term goals. For example, heroin users have been shown to undervalue a reward (money) when they had to wait for it. Heroin users also show impulsivity on pen-and-paper tests, such as completing a maze.

Essentially, addiction is an impulsive form of mood management. Someone chooses to get immediate results by taking a drug rather than going through long-term behavioral changes, even though they eventually are more effective.

Coping strategies can also protect against addiction or relapse. A study of heroin users found that those who used cognitive, avoidance, and distraction coping strategies were better able to avoid relapse.

ADDICTION IN THE BRAIN

A great deal of research is being done nowadays to delineate the neurobiology of addiction. All aspects of a person's mind and behavior are reflected in the activity of brain cells. The trick is to discover exactly which brain cells and what type of activity relate to addiction. Research in this field hopes to understand what changes in a person's brain when addiction occurs, as well as what may make some people more susceptible than others. For example, millions of people have tried cocaine at some point, but only some of them go on to use it regularly and become addicted. There must be some neurobiological difference between those two groups.

The brain has a **reward system,** which encourages us to pursue pleasurable activities and avoid aversive activities. This system is in place naturally for many purposes. It ensures that we satisfy our physiological needs, such as food, water, and sex. Not surprisingly, people take great pleasure in receiving these physical necessities. (Although sex is not necessary for survival the way food or water are, it is necessary for life to continue by procreation.) The reward system also adapts to accommodate abstract rewards such as money. Literally speaking, money is just printed paper, metal discs, or numbers on a bank statement, but through conditioning we learn that money can get us other things. The reward system of the brain has been shown to respond to both physiological and abstract rewards.

In the anatomy of the brain, the frontal lobes and limbic system are important parts of the reward system, and thus highly relevant to addiction. The frontal lobes are involved in executive functions, such as planning, decision making, motivation, regulating behavior, and impulse control. Any drug that has pleasurable effects is, in effect, shaping the activity of the frontal lobes to seek it out again in the future. The patterns of activity in the frontal lobes become focused on things that are rewarding, whether they are praise from a family member, a raise from an employer, or a line of cocaine. All behavior is a matter of making choices, and the frontal lobes are especially implicated in choice. For example, whenever someone decides to spend hard-earned cash on a drug (a pack of cigarettes, a beer, or a can of cola), it is done in place of buying food, a movie ticket, or saving it in a bank account. These choices are weighed in the frontal lobes.

The limbic system, as mentioned in previous chapters, represents emotional experiences in the brain. Accordingly, drugs that create pleasurable feelings and euphoria do so by altering activity in the limbic system. The limbic system and the frontal lobes communicate extensively with each other. In a sense, a mood-altering drug is tricking your brain. Emotional reactions typically occur in response to real-life events, such as getting a promotion, eating a delicious meal, or falling in love. We're designed to have those emotional reactions whenever something beneficial is happening to us. Taking a mood-altering drug is activating that same system, creating the emotional reaction, but without any real benefit. Thus, the brain is tricked into thinking that the drug-taking itself is a beneficial event even though the drug itself offers little or no practical benefit (and may instead be a financial, emotional, or physical drain). On some level, the addicted person is attributing significance or importance to the drug, which doesn't deserve importance since it doesn't offer any lasting benefit to the individual.

Studies have shown activation of the reward system in response to tasty foods such as chocolate, pleasant smells, a soft caress on the skin, enjoyable music, or even winning money. Brain imaging studies also show that taking a mood-lifting drug, such as nicotine, cocaine, and heroin, similarly activates the frontal lobes and limbic system, corresponding to the lift in mood they create. The reward system also responds to anything that is new or interesting, which helps explain why people like changes once in a while, to break from their usual routines (such as going on vacation). The reward system also activates in response to sexual arousal and orgasm. It's no surprise that love has been called a drug, since it has such powerful effects on the reward system.

In addition to the anatomy of the brain, we must also consider its chemistry. Several neurotransmitters are involved in reward and addiction. The one that has received the greatest attention is dopamine. Neurons that produce dopamine originate in the brainstem and project to higher areas of the brain, such as the frontal lobes and limbic system. Results from numerous experiments suggest that these dopamine neurons are active whenever one experiences something pleasurable or novel and interesting. These neurons release their dopamine in the frontal lobes and limbic system, focusing their activity on whatever caused the release of dopamine (e.g., food, sex, money, or a drug). Dopamine neurons even start to fire signals when a reward is anticipated (e.g., when waiting for your meal at a restaurant), but they become silent if a reward is not delivered (e.g., the food is really bad). The neurotransmitter serotonin has also been implicated in reward, particularly in the ability to tolerate a delay before getting a reward. Thus, low serotonin activity makes a person less able to delay gratification, or more impulsive.

So an addiction is essentially dysfunction of the reward system of the brain. It is learned, to some degree, as the mood-altering drug is taken repeatedly. The neuroscientist George Koob has defined addiction as **homeostatic hedonic dysregulation.** In essence, this suggests that humans maintain a regular range of pleasure in our lives. All of the enjoyable things we do keep us in a certain range of happiness, neither too high nor too low. Drugs that cause intense pleasure, such as cocaine or heroin, send our pleasure levels off the charts. After experiencing highs like that, normal reinforcers, such as a scoop of ice cream or a hug, don't seem so great anymore. The intense highs distort our reward economy, so that ordinary pleasures don't do the trick.

To a limited degree, susceptibility to addiction may be inherited. David E. Comings has defined a **reward deficiency syndrome.** He and his colleagues have done extensive research showing genetic differences in people with addictions. For example, genetic variations in a certain dopamine receptor make them under-sensitive to dopamine. It is presumed that this makes some people deficient in a sense of reward when they do something pleasurable. As a result, they must resort to more extreme measures to bring their pleasure levels up to par. Thus, they are more likely to develop addictive patterns, such as using drugs.

ARE ADDICTIONS LIMITED TO DRUGS?

Addictions are not limited to drugs by any means. Really, anything that creates pleasure could be used in an addictive pattern. One brain imaging study actually showed that playing a computer game releases dopamine in the brain. When the comedian Robin Williams said that Nintendo was "cocaine for kids," he wasn't far off the mark. Many activities similarly activate the reward system of the brain, and can be used maladaptively to manage mood.

The idea of behavioral addiction, or an addiction to an activity instead of a chemical, is fairly well accepted. Nothing about the concept of addiction requires that a drug be involved. Addictions to gambling, food, sex, spending money, exercise, and the Internet have been recognized. Potentially, anything that alters emotions in a pleasurable way (i.e., increasing pleasing emotions or decreasing unpleasant emotions) can be used in an addictive pattern.

Remember that the feature that tips the balance from use to abuse is cost versus benefit. To take gambling as an example, someone who gambles a small, set amount of money infrequently is doing so in an adaptive pattern. Someone who gambles large amounts of money compulsively is doing more harm than good. Many with a gambling addiction (also called "compulsive gamblers" or "pathological gamblers") admit that they gamble to alleviate stress, sadness, or anxiety, suggesting dependence on gambling to manage their mood. They even show tolerance: the excitement of smaller bets wears off over time, so that larger and larger stakes must be made to get the same boost.

The same parallels can be made with other behaviors such as eating. Bulimia and binge eating disorder are both conditions in which people eat massive amounts of food, far exceeding what they need for nutrition. Eating is pleasurable, and these conditions represent an immediate, impulsive means of getting pleasure. The long-term consequences of this type of eating are either obesity, eating restriction, or purging to avoid weight gain. Sexual addiction is particularly difficult to deal with since the means of getting pleasure (i.e., one's own genitals) are so readily available. Drug or gambling addicts can at least make efforts to avoid places that would tempt them.

Not surprisingly, people with one type of addiction have a higher chance of having another type of addiction as well. For example, the rate of drug addiction is higher than normal among people with gambling addiction, and vice versa. So it's somewhat short-sighted to think of a drug as the main problem in an addiction. It's more of an overall approach to managing pleasure, pain, anxiety, and stress.

Since all of the addictions have so many similarities, it's ironic that they can be regarded so differently. An overeater is seldom viewed in the same light as a heroin abuser, yet in many ways both suffer the same underlying problem. That probably has to do with the greater risks of a drug such as heroin, which has more serious, immediate risks including legal entanglements, financial loss, or contracting a disease (when the user injects it). Someone who overeats will likely suffer illnesses such as cardiovascular disease or diabetes eventually, so the potential for harm, if not fatality, is equally great. Addiction will only worsen the odds of untimely demise, regardless of the object of the addiction.

TREATMENT

Understanding addiction on multiple levels (biological, psychological, social, etc.) is essential to its treatment. It is hoped that improving our understanding will guide us to better treatment and prevention, since those are ultimately what matter.

Readiness for Treatment

The "stages of readiness" for treatment of addiction were developed by DiClemente and Prochaska (see Box 11.3) to help under-

BOX 11.3.
DiClemente and Prochaska's Stages of Readiness

1. Precontemplation
2. Contemplation
3. Determination
4. Action
5. Maintenance
6. Termination

stand where people are in terms of changing their drug use. People in the precontemplation stage are not yet thinking that their drug use (or other addiction) is harmful, so they don't see it as a problem. They may lack knowledge that the drug is actually doing them any harm, or they may think that people complaining about their drug use are overreacting. Other addicts may see an attempt at quitting as hopeless.

The next stage is contemplation, when people begin to acknowledge a problem and are considering the possibility of changing, but are still on the fence about it. At this stage, they may just be seeking information and exploring the possibilities. At this point, an addiction counselor would help them consider the risks versus rewards of using the drug.

The determination phase marks a decision to stop the drug use. Some mixed feelings may be lingering, but they're not holding the person back from making the attempt. It would be best to make a realistic plan at this point for how to stop using the drug, with concrete solutions to the problems that may arise. Simply deciding not to use the drug anymore is not a thorough plan. Withdrawal symptoms must be managed, and the user needs a strategy for dealing with the cravings that may occur, and for dealing with stress in other ways than getting drunk or high. If it were as simple as making a decision, few people would ever be addicted.

The action phase involves implementing the plan, which often begins with a public commitment such as attending Alcoholics Anonymous meetings and/or telling family and friends. This serves a few worthwhile purposes. It encourages a social support system to assist the person in quitting. It also allows people other than the one quitting to monitor the situation and watch for slips. The self-deceiving nature of an addiction makes it useful to rely on others (who one trusts) as a kind of reality-check system. Of course, using the right people (e.g., a sponsor or friend) can make a big difference. Nonetheless, a good plan will work over time, even if adjustments must be made. This kind of success is encouraging and reinforcing, helping the person maintain abstinence. Everyone is different, but the action phase often lasts three to six months.

Once success is met, the person is in a maintenance phase, since sustained abstinence is ultimately what matters. Maintenance can last for years, or maybe for the remainder of the person's life. As newer, more adaptive patterns are established, the risk of relapse becomes less severe. Unfortunately, the nature of the illness makes relapse always a possibility. Some lower their guard against the problem, or feel like testing themselves by engaging in a small degree of use. Others may meet with a great deal of difficulty or stress and not have the skills to cope with it. Fortunately, people at this stage are better equipped to quit again than a beginner, since they are already familiar with the problems and have the social supports, such as a sponsor. In the long run, people can benefit from their relapses, learning where the dangers are. As in most endeavors, persistence pays off.

The goal of addiction treatment is termination, when the person no longer sees the drug as a threat. At this point, the person has identified the situations that can trigger use. He or she has been able to deal with significant adversity by using coping and problem-solving skills, and not needing the drug. It has been debated whether anyone truly reaches this stage. Some believe that life after an addiction is indefinite maintenance of abstinence. Although the risk of relapse diminishes, abstinence is more of a process than a destination.

TREATMENT METHODS

Although substance abuse may be a stubborn problem, it is a treatable disorder.

Drug Treatment

It may seem ironic to treat a drug addiction by giving another drug, but in some cases it's beneficial or even necessary. Simply replacing one drug with another is obviously not a good strategy. Medications are used to alleviate the withdrawal syndrome caused by abstinence, particularly when it is too severe or even life-threatening. For example, heroin and other opioids cause a hellish withdrawal syndrome that can last up to seventy-two hours. It would be cruel and unnecessary to make people go through it. Instead, drugs such as methadone or LAAM (L-alpha-acetylmethadol) are used to gradually wean the person. This method is no picnic either, but at least it is more manageable. The drug clonidine also helps further reduce the withdrawal symptoms. In the case of alcohol or other sedatives, rapid withdrawal in a person who uses large amounts can be dangerous, causing delirium or even seizures. In those cases, they are switched to a safer drug, such as a benzodiazepine, and slowly weaned.

Another type of medication used in addiction treatment is anticraving drugs. These drugs dampen the urge to use the drug again, making it somewhat easier for the abstinent person to resist temptation. A drug in this category is not sufficient treatment by itself, but may help people in the early stages who have not yet learned to change their behavior and coping skills. Few effective anticraving drugs exist, but obviously they are of great interest. For example, naltrexone has shown some usefulness in reducing cravings for alcohol.

Medication is also used to help remove temptation. These drugs render the abused drug ineffective, or otherwise make taking it undesirable. For example, naltrexone nullifies the effects of opioids such as heroin. So people taking naltrexone daily (or receiving a skin implant of the drug) have less motivation to take heroin because they know it wouldn't have an effect if they took it anyway. Naltrexone by itself, however, doesn't seem to have any apparent psychological effects. In a similar vein, the drug Antabuse (disulfiram) makes people violently ill if they consume alcohol. Thus, it could be taken in the morning to discourage slips later in the day. Of course, these strategies assume that people are willing and cooperative.

Last, drugs that reduce depression or anxiety are sometimes used by people quitting an addiction. Although changes in behavior are a better long-term solution, depression or anxiety that is too severe can hinder progress. Antidepressant drugs (such as Zoloft, Paxil, Wellbutrin) are used temporarily for this purpose, since they don't have any abuse potential themselves. Again, it would be cruel and unnecessary to let a person languish in misery when an appropriate medication is available. In some cases, it may even help the recovery process. For example, people with depression experience even more severe withdrawal symptoms when quitting tobacco, and they have a greater risk of relapsing.

Psychotherapeutic Treatment

Although helpful in the right situation, a medication is never a sufficient treatment for an addiction. Psychotherapeutic treatment is necessary and comes in many forms. In the broadest sense, these treatments include individual counseling or psychotherapy and support groups (such as Alcoholics Anonymous). In some cases, such as heroin, a combination of medication and psychotherapeutic treatment gives the best odds.

Addiction treatment programs may operate on an outpatient or residential basis, and treatment may be short term (less than six months) or long term (six to twelve months). Outpatient treatment is best for people who have fairly stable, socially integrated lives and a shorter history of addiction. Short-term residential programs usually last three to six weeks (or the common twenty-eight days), and typically involve long-term follow-up with a group such as Alcoholics Anonymous or Narcotics Anonymous. Long-term residential treatment (six to twelve months) is used for people with a long history of drug dependence, criminal activity, and social dysfunction. These programs are very structured and attempt to resocialize the person into a better lifestyle.

Treatment is a long and expensive process, but worth every cent. Adding up the costs of medical care, crime, drug abuse treatment, social welfare programs, and time lost from work, drug addiction costs an estimated $67 billion per year. However, studies have shown that we save four to seven dollars for every dollar spent on treatment. Programs in special circumstances, such as prisons, are also worthwhile. In some programs, prisoners who get addiction treatment have a rearrest rate that is lower by one-quarter to one-half. Consider that it costs about $3,300 per month to incarcerate someone, but only $290 per month for methadone maintenance therapy.

PREVENTION

Although addiction treatment is necessary and beneficial, prevention is even better. The adage about an ounce of prevention being worth a pound of cure is almost literally true. In financial terms, it is estimated that every dollar spent on prevention saves a community four to five dollars in costs for treatment and counseling.

Years of research have identified factors that place a person at greater risk for addiction, as well as others that reduce the risk. Research funded by the National Institute on Drug Abuse (part of the National Institutes of Health) shows that prevention programs are effective. Factors that protect against drug abuse include strong family bonds, parental monitoring and involvement, clear and consistent discipline, and involvement with prosocial organizations (such as schools or clubs). Not surprisingly, risk factors include a chaotic home environment (including members with mental illness or addiction), ineffective parenting (e.g., inconsistent or nonexistent discipline), poor school performance, poor coping skills, and affiliation with deviant peers (i.e., hanging with the wrong crowd).

Naturally, prevention programs aim to foster factors that protect against drug abuse. Considering that addiction is a maladaptive form of mood management, it would make sense to teach adaptive forms particularly to children. Although many people learn adaptive skills from parents and family during their early years, many do not, or they learn maladaptive behaviors and the cycle is repeated. For this reason, we cannot rely entirely on families to fix their own problems. Community-based programs help in this regard, educating children as well as parents.

It is unfortunate that these types of programs are not integral parts of our educational system. Although subjects such as mathematics and reading are part of every grammar school curriculum, no widespread program teaches children coping strategies and problem-solving skills for their personal lives. However, when someone develops an addiction, we expect the person to attend treatment and quickly absorb all of these principles. That's equivalent to giving out driver's licenses and then requiring driver education for people who get into accidents. During the Reagan administration in the 1980s, the Just Say No campaign was launched by First Lady Nancy Reagan.

Although well intentioned, it demonstrates a tragic lack of understanding about the causes of addiction. Certainly, shaping attitudes may reduce drug use overall, but Just Say No doesn't teach children anything about what to do when they are angry, sad, or nervous. It doesn't prepare them to handle losing a job, paying bills, having a relationship, or losing a family member. Teaching skills in coping, communication, and problem solving is designed to do just that.

CONCLUSIONS

Addiction is a common and stubborn problem. A clear-cut definition may never be found, but most will agree that it is a self-destructive pattern of behavior that applies to drugs as well as many other behaviors. It can be fully understood only by examining it at biological, psychological, and social levels. Although relapse is a common problem and a simple cure does not seem to exist, treatment is available in the form of counseling and an increasing array of medications. Moreover, solid evidence exists to demonstrate the effectiveness of certain prevention programs. Treatment and prevention are not only humane approaches to addiction but also benefit society and are cost-effective. It is questionable whether addiction will ever be eradicated, but our continued efforts to improve the situation are worthwhile.

Glossary

acetylcholine: A neurotransmitter that is important in memory and the voluntary control of muscles.

action potential: The positive voltage of a neuron that is excited and transmitting a signal.

agonist: A drug that stimulates a receptor, imitating one of the neurotransmitters.

allspice: A plant *(Pimenta dioica)* used as a culinary spice that has local anesthetic effects.

Alzheimer's disease: The most common form of dementia, initially causing problems in memory and language.

amygdala: A part of the limbic system involved in the emotional significance of experiences.

analgesia: The relief of pain.

analgesia system: A system descending from the brainstem to the spinal cord that stops pain signals.

anandamide: The first discovered endocannabinoid neurotransmitter, which is produced in the brain and has effects similar to THC.

anesthesia: Local anesthesia refers to stopping all sensations by blocking transmission in a nerve. General anesthesia involves inducing deep sedation so that a patient is unconscious during painful or unpleasant procedures.

angel's trumpet: Plants (*Brugmansia* species) containing the drug scopolamine, which has hallucinogenic and memory-inhibiting effects.

antagonist: A drug that covers and blocks a receptor, preventing the effects of other neurotransmitters or drugs at that receptor.

antipsychotic medications: Drugs used to treat schizophrenia and other psychotic disorders by blocking receptors for the neurotransmitter dopamine in the brain. Also known as neuroleptics.

apigenin: Possibly the active chemical found in chamomile.

apolipoprotein E: A bodily chemical that has been implicated in Alzheimer's disease if a person has a certain variant.

areca: A palm tree *(Areca catechu)* that grows in tropical regions that produces a nut chewed for stimulant effects.

arecoline: The main stimulant drug in the areca nut, which works by stimulating receptors for the neurotransmitter acetylcholine.

aspirin: Acetylsalicylic acid, a drug used for analgesic, anti-inflammatory, anticlotting, and fever-reducing effects.

atypical antipsychotic drugs: A newer group of drugs that alleviate the symptoms of schizophrenia. In addition to reducing positive symptoms (e.g., hallucinations, delusions) as the traditional antipsychotic drugs did, they also alleviate some of the negative symptoms (e.g., social withdrawal). Traditional antipsychotic drugs block only dopamine receptors in the brain, while the atypical ones additionally block serotonin receptors.

autonomic nervous system: A set of nerves that control automatic functions in the body, such as digestion and heart rate.

axon: A long extension from the cell body of a neuron that carries signals toward the next neuron.

barbiturates: A group of drugs used for sedation including phenobarbital and pentobarbital.

basal ganglia: Structures deep in the brain that help control movements and thinking.

belladonna: A plant *(Atropa belladonna)* containing the drug scopolamine, which has hallucinogenic and memory-inhibiting effects.

benzodiazepines: A group of drugs used for sedation and the relief of anxiety, including Valium, Xanax, Rohypnol, and Librium.

beta-amyloid: A protein found in excess in the brains of people with Alzheimer's disease.

beta-carbolines: The active chemicals harmine and harmaline, which are found in *Banisteriopsis caapi,* used in the mixture known as yage.

biochemistry: The study of the chemistry of living organisms, such as those found in plants and animals.

biotransformation: The process of metabolizing a drug, inactivating it so that it may be eliminated more easily.

blood-brain barrier: A barrier that limits most chemicals, many potentially harmful, from entering the brain and spinal cord.

botany: A branch of biology that studies plants.

bradykinesia: Literally meaning "slow movement," this term refers to the slow movements caused by muscle rigidity in Parkinson's disease.

brainstem: The lower part of the brain in between the cerebrum and spinal cord. It controls many basic functions such as breathing and wakefulness.

caffeine: A stimulant found in coffee, tea, cocoa, guarana, maté, and kola nuts that works by blocking receptors of the neurotransmitter adenosine.

cannabinoids: A group of chemicals that are unique to cannabis plants. These include delta-9-tetrahydrocannabinol, cannabinol, and cannabidiol.

cannabis: A plant (*Cannabis sativa, C. indica,* and *C. ruderalis*) used for its fibers (hemp) and psychoactive effects for both recreational and therapeutic purposes.

capsaicin: The active chemical in chili peppers that, when applied externally, stimulates a heat sensor in the skin (the vanilloid receptor), which causes a person to feel a burning sensation.

carbidopa: A drug that enhances the effects of L-dopa, given to people with Parkinson's disease.

carbon monoxide: A toxic chemical that is actually used in very small amounts in the brain as a neurotransmitter.

cathinone: A stimulant drug from khat that works by increasing the amount of the neurotransmitters dopamine, norepinephrine, and serotonin in the synapse (by triggering their release).

catnip: A plant *(Nepeta cataria)* that has sedative effects in humans.

cell body: The main section of a cell that contains the nucleus and certain organelles. Synonymous with soma.

cerebellum: A structure located on the back of the brainstem that is involved in fine-tuning movements, making them smoother and more coordinated. More recent research indicates that the cerebellum is involved in fine-tuning thinking abilities as well.

cerebrospinal fluid (CSF): The fluid in and around the brain and spinal cord. It supports these structures and cushions them against impact.

chamomile: A daisylike plant (German chamomile, *Matricaria recutita,* and Roman chamomile, *Chamaemelum nobile*) used for many purposes, including sedative and anti-anxiety effects.

chili: A plant (*Capsicum annuum, C. frutescens,* and others) that produces a burning sensation when eaten or touched.

chrysin: The active chemical in passionflower, which works by enhancing the effects of the neurotransmitter GABA.

clove: A tree *(Syzygium aromaticum)* whose seeds are used as a culinary spice. An oil is extracted from the flowers that is used as a local anesthetic.

coca: A shrublike tree *(Erythroxylon coca)* containing cocaine, used as a stimulant or local anesthetic.

cocaine: A stimulant drug found in the coca plant that works by increasing the amount of the neurotransmitters dopamine, norepinephrine, and serotonin in the synapse (by blocking their reuptake).

cognitive: Relating to processes of thinking, including reasoning, abstract thinking, memory, and mental flexibility.

congenital insensitivity to pain: A disorder that renders a person insensitive to pain.

corpus callosum: Large band of axons used for communication between the left and right hemispheres of the brain.

cortex: (also cerebral cortex) The layer of gray matter covering the brain, which is divided into two hemispheres, each with four symmetrical lobes: frontal, parietal, occipital, and temporal.

cross-tolerance: The phenomenon where developing tolerance to one drug automatically creates tolerance to other similar drugs (e.g., cocaine and amphetamine).

cyclooxygenase: An enzyme that converts prostaglandins, which sensitize pain nerves. Drugs such as aspirin block cyclooxygenase, which prevents pain nerves from becoming more sensitive.

dementia: A collective term for several illnesses that cause a progressive loss of higher cognitive functions.

dendrite: An extension branching from the body of a neuron, primarily forming the end of the neuron that receives signals.

dependence: Requiring a drug to continue functioning because stopping would cause a withdrawal syndrome.

depressive pseudodementia: A condition occurring in some depressed elderly people that resembles Alzheimer's disease. However, it is entirely due to depression, so the person can improve to normal status when the depression is treated.

doctrine of signatures: Belief that the shape or color of a plant indicates its therapeutic uses.

dopamine: A neurotransmitter involved in thought, emotion, appetite, sex, and perception. It is especially important in reward and addiction.

dose-response curve: A graph that demonstrates the relationship between the amount of the drug taken and the magnitude of effect.

dronabinol: A synthetic form of THC taken orally for medical use.

drug: A nonnutritive chemical that affects bodily processes, often used to treat an illness or for recreational purposes.

drug absorption: The process of a drug entering the body, through the gastrointestinal tract (enteral) or otherwise (parenteral).

dyskinesia: Abnormal movements that can develop with high doses of medications for Parkinson's disease.

effective dose: The amount of a drug needed to create the desired effect.

efficacy: The effectiveness of a drug. Drugs with more intense effects are said to have greater efficacy (e.g., morphine has greater efficacy than aspirin for relief of pain).

electroencephalogram (EEG): A means of measuring brain activity through electrical voltage produced by neurons as they send signals. Beta activity indicates an awake and alert state; alpha activity indicates an awake but inactive or resting state; theta activity indicates sleep; and delta activity indicates deep sleep. Beta activity also occurs paradoxically when a person is in REM sleep, indicating that the brain is very active while dreaming.

elemicin: One of the active chemicals in nutmeg and mace, having hallucinogenic effects and similar in structure to mescaline.

endocannabinoids: A group of neurotransmitters discovered because of research done on cannabis. They play normal roles in memory, thinking, movement, and pain perception.

endorphin: A neurotransmitter involved in many functions, including control of pain.

enteral: Administration of a drug through the gastrointestinal tract, including sublingual, oral, and rectal administration.

ephedra: A shrub *(Ephedra sinica)* that is used for stimulant effects.

ephedrine: The main stimulant drug in ephedra that works by stimulating receptors for the neurotransmitter norepinephrine.

ergot: A fungus *(Claviceps purpurea)* found in grains. It contains a hallucinogenic chemical closely related to LSD and is responsible for the unpleasant symptoms of ergotism, or St. Anthony's fire.

ethnobotany/ethnopharmacology: The study of the use of plants and drugs in the context of culture.

eugenol: A chemical found in clove and allspice that has local anesthetic effects. It works by preventing electrical transmission in nerves, similarly to local anesthetics used by dentists and physicians.

evolutionary mismatch: The concept that, due to civilization (e.g., technology, agriculture), humans currently live in circumstances that are very different from those of our ancestors, who were nomadic hunter-gatherers. Many health and behavioral problems arise because evolution has shaped us to be successful in our earlier context, and not to deal with many modern situations (e.g., unlimited food supply, potent psychoactive drugs).

excretion: The process of removing a drug or its breakdown products from the body through urine, feces, sweat, saliva, tears, hair, and/or exhalation.

fava: A bean *(Vicia faba)* that contains high amounts of L-dopa, which can interfere with medications for Parkinson's disease.

feverfew: A plant *(Tanacetum parthenium)* traditionally used to treat pain, migraine, arthritis, fevers, and anxiety.

flavonoid glycosides: A class of active chemicals found in ginkgo.

fly agaric: A mushroom *(Amanita muscaria)* that has hallucinogenic effects and may be the oldest and most widely used hallucinogen. It contains several chemicals that may have hallucinogenic effects (such as ibotenic acid, muscimol, muscazone, and muscarine), which work by interfering with the neurotransmitters GABA and glutamate.

frontal lobe: Portion of the cerebral cortex involved in voluntary movement, language production, and executive functions (e.g., planning, organization, flexibility, self-control).

furanosesquiterpenes: Active chemicals in the plant myrrh, which create analgesic effects.

GABA: A neurotransmitter that has strong inhibitory effects on neurons.

ginger: A plant *(Zingiber officinale)* whose rhizome (underground stem) is used as a culinary spice and for antinausea effects. It may have some mild anti-anxiety and analgesic effects.

gingerol: One of the active chemicals in ginger.

ginkgo: A tree *(Ginkgo biloba)* whose leaves are consumed to improve memory, among other uses.

ginseng: A plant *(Panax ginseng)* whose rhizome (underground stem) is used for multiple purposes, including memory improvement.

ginsenosides: The active chemicals found in ginseng.

glial cells: Fatty support cells of the nervous system that help neurons survive and function.

glutamate: An amino acid that is used as a neurotransmitter. It has strong excitatory effects on neurons.

hallucinogen: A drug that has profound psychological effects, producing false perceptions (i.e., hallucinations) but also affecting cognition and emotion.

hallucinogen-persisting perception disorder (HPPD): A condition in which the effects of hallucinogenic drugs are reexperienced, even though the drug has not been taken again. A common term for this phenomenon is "flashbacks." It occurs in a portion of hallucinogen users, and some cases may involve other neurological or psychiatric illnesses with symptoms that resemble HPPD.

henbane: A plant *(Hyoscyamus niger)* containing the drug scopolamine, which has hallucinogenic and memory-inhibiting effects.

heroin: Street name for the drug diacetylmorphine, which has intense emotional effects and high abuse potential.

hippocampus: A part of the limbic system that is essential for making new memories.

homeostatic hedonic dysregulation: A concept originated by George Koob that certain drugs have such powerful effects that they distort our reward economy, so that normal pleasures no longer satisfy.

hops: A flowering vine *(Humulus lupulus)* that has sedative effects. It is also used to give beer a bitter flavor.

hormones: Chemical messengers that travel throughout the body and sometimes the brain, carrying out a variety of functions. They include sex hormones (e.g., testosterone, estrogen, and progesterone), stress hormones (e.g., cortisol), and thyroid hormones (e.g., thyroxine).

hyoscyamine: A hallucinogenic drug found in several plants that works by blocking muscarinic receptors for the neurotransmitter acetylcholine.

hyperalgesia: A pain disorder involving an excessive sensitivity to pain.

hyperforin: A chemical in St. John's wort that has antidepressant effects because it increases active levels of dopamine, serotonin, and norepinephrine in the brain.

hypericin: A chemical in St. John's wort that acts as a weak MAO inhibitor and has possible antidepressant effects in combination with other chemicals.

hypothalamus: A structure in the brain that monitors important physiological functions (sleep, appetite, thirst, sex drive, and body temperature), coordinating them with behavior, hormones, and emotions.

iboga: A shrub *(Tabernanthe iboga)* from west central Africa that has hallucinogenic and potentially anti-addictive effects. It contains several active chemicals, including ibogaine.

ibogaine: One of the active chemicals of iboga, which works by stimulating serotonin receptors in the brain, similar to LSD. It also blocks the neurotransmitter glutamate, which may relate to its anti-addictive effects.

intramuscular: Administering a drug by injecting it into a muscle.

intravenous: Administering a drug by injecting it into a vein.

kava: (also known as kava kava) A plant *(Piper methysticum)* native to the South Pacific islands, taken for sedative and anti-anxiety effects.

kavalactones: A group of chemicals that cause the sedative and anti-anxiety effects of kava.

ketamine: A synthetic hallucinogen, also known colloquially as Special K, originally developed by a pharmaceutical company as a general anesthetic for surgery.

khat: A shrub *(Catha edulis)* or tree native to East Africa whose leaves are used for stimulant effects.

LAAM: L-alpha-acetylmethadol, a drug used to treat opioid addiction.

L-dopa: The precursor for the neurotransmitter dopamine, used as a treatment for Parkinson's disease.

lemon balm: A plant *(Melissa officinalis)* with potential sedative effects.

lethal dose: The amount of a drug that will kill a person who takes it.

limbic system: A set of structures concerned with emotions, including the hippocampus, amygdala, and other structures.

lobelia: A flowering herb that is used for stimulant effects.

lobeline: The main stimulant drug in lobelia that works by stimulating receptors for the neurotransmitter acetylcholine.

LSD: Lysergic acid diethylamide, a hallucinogenic drug inadvertently created by Albert Hofmann in 1949. It works by stimulating certain serotonin receptors in the brain.

mace: A culinary spice from the seed covering of the tree *Myristica fragrans,* which also has hallucinogenic effects if taken in sufficient quantity.

mandrake: A plant *(Mandragora officinarum)* containing the drug scopolamine, which has hallucinogenic and memory-inhibiting effects.

MAO: Monoamine oxidase, an enzyme that breaks down the neurotransmitters dopamine, norepinephrine, and serotonin.

MAO inhibitors: Antidepressant drugs that block monoamine oxidase, increasing levels of the neurotransmitters dopamine, norepinephrine, and serotonin in the brain.

meadowsweet: A plant *(Filipendula ulmaria)* that produces salicylate, which is closely related to the precursor of aspirin.

meninges: Protective layers covering the brain and spinal cord.

mescaline: The active hallucinogenic chemical in the peyote cactus. Like LSD, it works by stimulating certain serotonin receptors in the brain.

methadone: A drug used to treat opioid addiction.

mitochondria: A component of cells that creates energy by converting glucose to ATP.

monoamines: A group of neurotransmitters including serotonin, dopamine, and norepinephrine.

morning glory: A group of flowering vines (e.g., ololiuqui, *Turbina corymbosa*), whose seeds have hallucinogenic effects due to the presence of LSD-like chemicals.

myelin: A fatty covering on the axons of some neurons that serves as insulation.

myristicin: One of the active chemicals in nutmeg and mace, which has hallucinogenic effects and is similar in structure to mescaline.

myrrh: A shrub *(Commiphora molmol)* from the Mediterranean and Middle East regions with potential analgesic effects.

naltrexone: A drug that blocks opioid receptors, used to reverse an overdose.

neuritic plaques: An accumulation of proteins (beta-amyloid and apolipoprotein E) found in the brains of people with Alzheimer's disease.

neurofibrillary tangles: A tangle of protein found in excess in the brains of people with Alzheimer's disease.

neurology: Medical discipline that diagnoses and treats illnesses of the nervous system.

neuron: A type of cell in the nervous system that processes information. The brain contains approximately 100 billion of them. They are highly interconnected into networks and pass messages by electrical and chemical signals.

neuropsychology: Branch of psychology that studies the nervous system and its relationship to mental processes such as thought and emotion.

neuroscience: The scientific study of the nervous system.

neurotransmitter: A chemical messenger released by neurons that can excite or inhibit other neurons.

nicotine: A stimulant drug in the tobacco plant that works by activating receptors for the neurotransmitter acetylcholine.

nitric oxide (NO): A toxic chemical that is used in very small amounts in the brain as a neurotransmitter.

nocebo: Adverse side effects that occur as a result of one's expectations rather than a physical effect of a drug or treatment (from Latin, meaning, "I shall harm").

nonsteroidal anti-inflammatory drugs (NSAIDs): A class of drugs used for analgesic, anti-inflammatory, anticlotting, and/or fever-reducing effects. They include aspirin, acetaminophen, and ibuprofen.

norepinephrine: A neurotransmitter involved in thought, emotion, appetite, sex, and perception.

nucleus: A component of cells that stores the genetic blueprint, DNA (deoxyribonucleic acid).

nutmeg: A culinary spice from the seed of a tree *(Myristica fragrans)* that has hallucinogenic effects if taken in sufficient quantity.

nutraceutical: A fortified food or dietary supplement (from the words *nutritive* and *pharmaceutical*), often blurring the distinction between food and drugs.

nutrient: A chemical that is necessary for the body to maintain normal health and growth.

occipital lobe: Portion of the cerebral cortex involved in vision.

opioid: Any opiumlike drug, whether natural or synthetic.

opium poppy: A plant *(Papaver somniferum)* that provides seeds used for baking, and the drugs morphine and codeine, which are used to treat pain and cough.

oxytocin: A neurotransmitter and hormone that is involved in social and emotional functions such as sex and orgasms.

pain asymbolia: A disorder that causes indifference to pain, even though a person is capable of feeling it.

parasympathetic nervous system: A branch of the autonomic nervous system that prepares the body for rest (e.g., by decreasing heart rate).

parenteral: Administration of a drug to a part of the body outside of the gastrointestinal tract. This would include any form of injection, snorting into the nose, or absorption through the skin.

parietal lobe: Portion of the cerebral cortex involved in perception of bodily sensations and body image.

Parkinson's disease: A degenerative brain disease that causes problems with movement, thinking, and depression.

parthenolide: The main active chemical in feverfew, which has aspirin-like effects.

passionflower: A plant (*Passiflora incarnata, P. coerulea,* and *P. edulis*) used for its sedative and anti-anxiety effects.

PCP: A synthetic hallucinogen, also known colloquially as angel dust, originally developed by a pharmaceutical company as a general anesthetic for surgery.

peyote: A spineless cactus *(Lophophora williamsii)* growing in Mexico and the Rio Grande valley, containing the hallucinogenic drug mescaline.

pharmaceutical: Pertaining to the manufacture and sale of drugs.

pharmacology: The study of drugs.

placebo: Beneficial effects that occur as a result of one's expectations rather than a physical effect of a drug or treatment (from Latin, meaning "I shall please").

potency: The relative amount of drug needed to produce a desired response. For example, if two drugs create equal effects, the one that does so at a lower dose is said to be more potent.

prostaglandin E: One of the body's chemical messengers that makes pain nerves more sensitive.

psilocin: The active hallucinogenic chemical in psilocybe and related mushrooms.

psilocybe: A group of mushrooms with hallucinogenic effects.

psilocybin: The active hallucinogenic chemical in psilocybe and related mushrooms.

psychiatry: Medical discipline that diagnoses and treats mental illnesses.

psychoactive: Having mind-altering effects, altering thought, memory, emotion, sensory perceptions, or body movement. Roughly synonymous with the term *psychotropic.*

psychology: Research and clinical discipline that studies the mind, brain, and behavior.

psychopharmacology: The study of psychoactive drugs.

psychotomimetic: A term that literally means "psychosis imitating." It has been applied to hallucinogenic drugs but is not accurate for many types (e.g., LSD) because the experiences are different.

receptor: A protein on a neuron that receives messages (in the form of neurotransmitters) from other neurons. Neurotransmitters fit into specific receptors like keys fit into locks.

reinforcement: The behavioral principle that we tend to repeat behaviors followed by positive, pleasurable consequences and not to repeat behavior followed by negative, aversive consequences.

REM sleep: A stage of sleep in which the brain is very active, showing beta waves that resemble conscious activity. REM stands for "rapid eye movement," referring to the quick eye movements that are a hallmark of this stage of sleep. Subjectively, REM sleep is associated with vivid dreams and nightmares.

reserpine: The main active chemical in serpentwood, which works by lowering levels of norepinephrine, serotonin, and dopamine in the nervous system.

resting potential: The negative voltage of a neuron that is at rest.

reticular formation: Part of the brainstem that keeps the brain activated when a person is alert and awake.

reward deficiency syndrome: A concept defined by David E. Comings suggesting that some individuals have genetic predispositions that make their reward systems less sensitive. Consequently, these individuals tend to seek more extreme forms of stimulation, which sometimes result in various addictions.

reward system: A system in the brain that encourages us to pursue pleasurable activities and avoid aversive ones, ensuring that we meet our basic needs, such as food, water, and sex. It involves multiple structures (e.g., the frontal lobes and limbic system) and neurotransmitters (e.g., dopamine).

Reye's syndrome: A rare but potentially fatal reaction to aspirin during a viral infection.

route of administration: The means by which a drug is administered, which can greatly influence the intensity of effect since it can alter the rate at which a drug is introduced and how much of it reaches the intended body part in active form.

salicylate: A chemical with analgesic, anti-inflammatory, anticlotting, and fever-reducing effects.

schizophrenia: A mental illness characterized by hallucinations, delusions, and other impairments in thinking.

scopolamine: A hallucinogenic drug found in several plants that works by blocking muscarinic receptors for the neurotransmitter acetylcholine.

scopolia: A plant *(Scopolia carniolica)* containing the drug scopolamine, which has hallucinogenic and memory-inhibiting effects.

serotonin: A neurotransmitter involved in thought, emotion, appetite, sex, and perception.

serpentwood: A small tree or shrub *(Rauwolfia serpentina)* whose root has been used historically for the treatment of schizophrenia in India. It has also yielded blood pressure medications.

shogaol: One of the active chemicals in ginger.

skullcap: A plant *(Scutellaria lateriflora)* with sedative effects.

solanaceous vegetables: A group of vegetables (tomato, potato, and eggplant) containing the chemicals alpha-solanine and alpha-chaconine, which could interfere with some medications, such as those prescribed for Alzheimer's disease (e.g., Aricept, Exelon, or Reminyl).

soma: The main section of a cell, which contains the nucleus and certain organelles. Synonymous with *cell body.*

spinal cord: The main route of input and output to the brain, containing nerves that go to and from the rest of the body. It also controls many motor reflexes.

SSRI: Selective serotonin reuptake inhibitors, a class of antidepressant drugs that increase active levels of the neurotransmitter serotonin.

St. John's wort: A plant *(Hypericum perforatum)* used for its antidepressant effects.

stimulant: A drug that causes mental alertness, reduces fatigue, and elevates mood.

subcutaneous: Administration of a drug by injecting it underneath a pinch of skin but not into muscle. In the case of heroin use, this is referred to colloquially as "skin popping."

sublingual: Administration of a drug by allowing it to absorb through the skin under the tongue.

substance abuse: A maladaptive pattern of drug use leading to physical, emotional, and social dysfunction of the individual.

substance dependence: A pattern of substance abuse over time, typically with an escalating pattern of use (i.e., tolerance), and causing a withdrawal syndrome if the person attempts to cut down or quit (i.e., dependence). Ultimately the long-term risks and costs of using the substance greatly outweigh any possible benefits.

substance P: A neurotransmitter involved in many functions, including sending pain signals to the brain.

substantia nigra: The part of the brain that degenerates in Parkinson's disease, depriving other parts of the brain of the neurotransmitter dopamine.

sympathetic nervous system: A branch of the autonomic nervous system that prepares the body for action (e.g., increasing heart rate).

synapse: The gap between neurons where communication occurs when one neuron releases a neurotransmitter and the other receives it.

synergy: Combining certain drugs can produce effects that are greater than the sum of their individual effects.

tardive dyskinesia: Involuntary, repetitive movements that occur in a portion of people who use antipsychotic drugs. In some cases, the symptoms can become permanent.

temporal lobe: Portion of the cerebral cortex involved in hearing and language comprehension.

terpene lactones: A class of active chemicals found in ginkgo.

thalamus: A structure in the brain with many functions. It serves as a relay station for the senses, and also helps keep the brain activated when awake.

THC: Delta-9-tetrahydrocannabinol. The main active chemical in cannabis. It works by acting on receptors in the brain for a newly discovered group of neurotransmitters called endocannabinoids (e.g., 2-AG, anandamide).

thorn apple: A plant *(Datura stramonium)* containing the drug scopolamine, which has hallucinogenic and memory-inhibiting effects.

tolerance: Taking a drug repeatedly leads to a diminished effect, requiring higher doses to sustain the same level of effect.

topical administration: Administering a drug by applying it to the skin for absorption (i.e., use of a cream or patch). It is synonymous with TRANSDERMAL ADMINISTRATION.

transdermal administration: Administering a drug by applying it to the skin for absorption (i.e., use of a cream or patch). It is synonymous with TOPICAL ADMINISTRATION.

2-AG: 2-arachidonylglycerol. An endocannabinoid neurotransmitter produced in the brain, which has effects similar to THC.

valerian: A plant *(Valeriana officinalis)* used for its sedative and anxiety-reducing effects.

vanilloid receptor: A receptor in the body that signals heat and pain. It is activated by the chemical capsaicin, found in chili peppers.

vascular dementia: A loss of cognitive function that results from compromised blood flow to the brain.

ventricles: A set of fluid-filled chambers in the brain, producing cerebrospinal fluid.

willow: A tree *(Salix nigra)* whose bark contains salicylate and has analgesic effects.

withdrawal syndrome: A constellation of symptoms that occurs when a drug is abruptly stopped, particularly when one has used it for a long time and/or at high doses. The symptoms of withdrawal typically are the opposite of those effects induced by the drug; e.g., withdrawal from a sedative would cause insomnia.

yage: A mix of hallucinogenic plants concocted in the Amazon, including the vine *Banisteriopsis caapi*.

Bibliography

Chapter 1

Cupp MJ (1999). Herbal remedies: Adverse effects and drug interactions. *Am Fam Physician* 59(5): 1239-1245.

Darwin C ([1859] 1988). *On the origin of species.* New York: New York University Press.

Dietary Supplement Health and Education Act of 1994 (1994). Pub L No. 103-417.

Eisenberg DM (1997). Advising patients who seek alternative medical therapies. *Ann Intern Med* 127(1): 61-69.

Eisenberg DM, Davis RB, Ettner SL, Appel S, Wilkey S, Van Rompay M, and Kessler RC (1998). Trends in alternative medicine use in the United States, 1990-1997: Results of a follow-up national survey. *JAMA* 280(18): 1569-1575.

Eisenberg DM, Kessler RC, Foster C, Norlock FE, Calkins DR, and Delbanco TL (1993). Unconventional medicine in the United States. Prevalence, costs, and patterns of use. *N Engl J Med* 328(4): 246-252.

Hadley SK and Petry JJ (1999). Medicinal herbs: A primer for primary care. *Hosp Pract (Off Ed)* 34(6): 105-106, 109-112, 115-116.

Kaptchuk TJ and Eisenberg DM (1998). The persuasive appeal of alternative medicine. *Ann Intern Med* 129(12): 1061-1065.

O'Hara M, Kiefer D, Farrell K, and Kemper K (1998). A review of 12 commonly used medicinal herbs. *Arch Fam Med* 7(6): 523-536.

Onopa J (1999). Complementary and alternative medicine (CAM): A review for the primary care physician. *Hawaii Med J* 58(2): 9-19.

Page JE, Balza F, Nishida T, and Towers GH (1992). Biologically active diterpenes from *Aspilia mossambicensis,* a chimpanzee medicinal plant. *Phytochemistry* 31(10): 3437-3439.

Robbers JE, Speedie MK, and Tyler VE (1996). *Pharmacognosy and pharmacobiotechnology.* Baltimore: Williams and Wilkins.

Rodriguez E, Aregullin M, Nishida T, Uehara S, Wrangham R, Abramowski Z, Finlayson A, and Towers GH (1985). Thiarubrine A, a bioactive constituent of *Aspilia* (Asteraceae) consumed by wild chimpanzees. *Experientia* 41(3): 419-420.

Siahpush M (1999). Why do people favour alternative medicine? *Aust N Z J Public Health* 23(3): 266-271.

Sixma HJ, Spreeuwenberg PMM, and van der Pasch MAA (1998). Patient satisfaction with the general practitioner: A two-level analysis. *Medical Care* 36(2): 212-229.

Southwood TR (1984). Insect-plant adaptations. *Ciba Found Symp* 102: 138-151.

Tyler V (1994). *Herbs of choice.* Binghamton, NY: The Haworth Press.

Wetzel MS, Eisenberg DM, and Kaptchuk TJ (1998). Courses involving complementary and alternative medicine at US medical schools. *JAMA* 280(9): 784-787.

Chapter 2

Cooper JR, Bloom FE, and Roth RH (1996). *The biochemical basis of neuropharmacology,* Seventh edition. New York: Oxford University Press.

DeArmond SJ, Fusco MM, and Dewey MM (1989). *Structure of the human brain,* Third edition. New York: Oxford University Press.

Hall ZW (1992). *Introduction to molecular neurobiology.* Sunderland, MA: Sinauer.

Hille B (1992). *Ionic channels of excitable membranes,* Second edition. Sunderland, MA: Sinauer Associates.

Kandel E, Schwartz JH, and Jessel TM (Eds.) (1991). *Principles of neural science,* Third edition. Norwalk, CT: Appleton & Lange.

Kebabian JW and Neumeyer JL (1994). *The RBI handbook of receptor classification.* Natick, MA: Research Biochemicals International.

Parent A (1996). *Carpenter's human neuroanatomy,* Ninth edition. Baltimore: Williams and Wilkins.

Siegel GJ, Agranoff BW, Albers RW, and Molinoff PB (Eds.) (1994). *Basic neurochemistry: Molecular, cellular, and medical aspects,* Sixth edition. New York: Lipincott-Raven.

Chapter 3

Clark WG, Brater DC, and Johnson AR ([1988] 1992). *Goth's medical pharmacology,* Twelfth edition. St. Louis: Mosby.

Cooper JR, Bloom FE, and Roth RH (1996). *The biochemical basis of neuropharmacology,* Seventh edition. New York: Oxford University Press.

Hardman JG, Gilman AG, and Limbird LE (1996). *Goodman and Gilman's the pharmacological basis of therapeutics,* Ninth edition. New York: McGraw-Hill.

Hille B (1992). *Ionic channels of excitable membranes,* Second edition. Sunderland, MA: Sinauer Associates.

Chapter 4

Alm B, Milerad J, Wennergren G, Skjaerven R, Oyen N, Norvenius G, Daltveit AK, Helweg-Larsen K, Markestad T, and Irgens LM (1998). A case-control study of smoking and sudden infant death syndrome in the Scandinavian countries, 1992 to 1995. The Nordic Epidemiological SIDS Study. *Arch Dis Child* 78(4): 329-334.

Arnold ME, Petros TV, Beckwith BE, Coons G, and Gorman N (1987). The effects of caffeine, impulsivity, and sex on memory for word lists. *Physiol Behav* 41(1): 25-30.

Asthana S, Raffaele KC, Greig NH, Berardi A, Morris PP, Schapiro MB, Rapoport SI, Blackman MR, and Soncrant TT (1995). Neuroendocrine responses to intravenous infusion of arecoline in patients with Alzheimer's disease. *Psychoneuroendocrinology* 20(6): 623-636.

Baer RA (1987). Effects of caffeine on classroom behavior, sustained attention, and a memory task in preschool children. *J Appl Behav Anal* 20(3): 225-234.

Balfour DJ (1994). Neural mechanisms underlying nicotine dependence. *Addiction* 89(11): 1419-1423.

Balfour DJ, Benwell ME, Birrell CE, Kelly RJ, and Al-Aloul M (1998). Sensitization of the mesoaccumbens dopamine response to nicotine. *Pharmacol Biochem Behav* 59(4): 1021-1030.

Barr HM and Streissguth A (1991). Caffeine use during pregnancy and child outcome: A 7-year prospective study. *Neurotoxicol Teratol* 13(4): 441-448.

Barry M (1991). The influence of the U.S. tobacco industry on the health, economy, and environment of developing countries. *N Engl J Med* 324(13): 917-920.

Belej T, Manji D, Sioutis S, Barros HM, and Nobrega JN (1996). Changes in serotonin and norepinephrine uptake sites after chronic cocaine: Pre- vs. post-withdrawal effects. *Brain Res* 736(1-2): 287-296.

Benwell ME and Balfour DJ (1998). The influence of lobeline on nucleus accumbens dopamine and locomotor responses to nicotine in nicotine-pretreated rats. *Br J Pharmacol* 125(6): 1115-1119.

Berlin I, Said S, Spreux-Varoquaux O, Launay JM, Olivares R, Millet V, Lecrubier Y, and Puech AJ (1995). A reversible monoamine oxidase A inhibitor (moclobemide) facilitates smoking cessation and abstinence in heavy, dependent smokers. *Clin Pharmacol Ther* 58(4): 444-452.

Bernstein GA, Carroll ME, Crosby RD, Perwien AR, Go FS, and Benowitz NL (1994). Caffeine effects on learning, performance, and anxiety in normal school-age children. *J Am Acad Child Adolesc Psychiatry* 33(3): 407-415.

Berry J, van Gorp WG, Herzberg DS, Hinkin C, Boone K, Steinman L, and Wilkins JN (1993). Neuropsychological deficits in abstinent cocaine abusers: Preliminary findings after two weeks of abstinence. *Drug Alcohol Depend* 32(3): 231-237.

Blumenthal M and King P (1997). The agony of the ecstasy. *HerbalGram* 37: 20-49.

Bolla KI, Cadet JL, and London ED (1998). The neuropsychiatry of chronic cocaine abuse. *J Neuropsychiatry Clin Neurosci* 10(3): 280-289.

Brenneisen R, Fisch HU, Koelbing U, Geisshüsler S, and Kalix P (1990). Amphetamine-like effects in humans of the khat alkaloid cathinone. *Br J Clin Pharmacol* 30(6): 825-828.

Briggs CA and McKenna DG (1998). Activation and inhibition of the human alpha7 nicotinic acetylcholine receptor by agonists. *Neuropharmacology* 37(9): 1095-1102.

Bryant CA, Farmer A, Tiplady B, Keating J, Sherwood R, Swift CG, and Jackson SH (1998). Psychomotor performance: Investigating the dose-response relationship for caffeine and theophylline in elderly volunteers. *Eur J Clin Pharmacol* 54(4): 309-313.

Bunker ML and McWilliams M (1979). Caffeine content of common beverages. *J Am Diet Assoc* 74(1): 28-32.

Cameron OG, Modell JG, and Hariharan M (1990). Caffeine and human cerebral blood flow: A positron emission tomography study. *Life Sci* 47(13): 1141-1146.

Carmody TP (1989). Affect regulation, nicotine addiction, and smoking cessation. *J Psychoactive Drugs* 21(3): 331-342.

Chiang WT, Yang CC, Deng JF, and Bullard M (1998). Cardiac arrhythmia and betel nut chewing—Is there a causal effect? *Vet Hum Toxicol* 40(5): 287-289.

Chu NS (1994). Effects of betel chewing on electroencephalographic activity: Spectral analysis and topographic mapping. *J Formos Med Assoc* 93(2): 167-169.

Clarke PB (1990). Mesolimbic dopamine activation—the key to nicotine reinforcement? *Ciba Found Symp* 152: 153-162; discussion 162-168.

Cone EJ (1995). Pharmacokinetics and pharmacodynamics of cocaine. *J Anal Toxicol* 19(6): 459-478.

Dackis C and Gold MS (1985). Neurotransmitter and neuroendocrine abnormalities associated with cocaine use. *Psychiatr Med* 3(4): 461-483.

Dager SR, Layton ME, Strauss W, Richards TL, Heide A, Friedman SD, Artru AA, Hayes CE, and Posse S (1999). Human brain metabolic response to caffeine and the effects of tolerance. *Am J Psychiatry* 156(2): 229-237.

Damaj MI, Patrick GS, Creasy KR, and Martin BR (1997). Pharmacology of lobeline, a nicotinic receptor ligand. *J Pharmacol Exp Ther* 282(1): 410-419.

D'Ambrosio SM (1994). Evaluation of the genotoxicity data on caffeine. *Regul Toxicol Pharmacol* 19(3): 243-281.

Decker MW, Brioni JD, Sullivan JP, Buckley MJ, Radek RJ, Raszkiewicz JL, Kang CH, Decker MW, Majchrzak MJ, and Arneric SP (1993). Effects of lobeline, a nicotinic receptor agonist, on learning and memory. *Pharmacol Biochem Behav* 45(3): 571-576.

Dodd SL, Herb RA, and Powers SK (1993). Caffeine and exercise performance: An update. *Sports Med* 15(1): 14-23.

Easton C and Bauer LO (1997). Neuropsychological differences between alcohol-dependent and cocaine-dependent patients with or without problematic drinking. *Psychiatry Res* 71(2): 97-103.

Ehlers CL, Somes C, Thomas J, and Riley EP (1997). Effects of neonatal exposure to nicotine on electrophysiological parameters in adult rats. *Pharmacol Biochem Behav* 58(3): 713-720.

Elmi AS (1983). The chewing of khat in Somalia. *J Ethnopharmacol* 8(2): 163-176.

Epping-Jordan MP, Watkins SS, Koob GF, and Markou A (1998). Dramatic decreases in brain reward function during nicotine withdrawal. *Nature* 393(6680): 76-79.

Foltin RW, Fischman MW, and Levin FR (1995). Cardiovascular effects of cocaine in humans: Laboratory studies. *Drug Alcohol Depend* 37(3): 193-210.

Ford RP, Schluter PJ, Mitchell EA, Taylor BJ, Scragg R, and Stewart AW (1998). Heavy caffeine intake in pregnancy and sudden infant death syndrome: New Zealand Cot Death Study Group. *Arch Dis Child* 78(1): 9-13.

Foulds J, McSorley K, Sneddon J, Feyerabend C, Jarvis MJ, and Russell MA (1994). Effect of subcutaneous nicotine injections of EEG alpha frequency in non-smokers: A placebo-controlled pilot study. *Psychopharmacology (Berl)* 115(1-2): 163-166.

Fowler JS, Volkow ND, Wang GJ, Pappas N, Logan J, MacGregor R, Alexoff D, Wolf AP, Warner D, Cilento R, and Zezulkova I (1998). Neuropharmacological actions of cigarette smoke: Brain monoamine oxidase B (MAO B) inhibition. *J Addict Dis* 17(1): 23-34.

Fox BS, Kantak KM, Edwards MA, Black KM, Bollinger BK, Botka AJ, French TL, Thompson TL, Schad VC, Greenstein JL, Gefter ML, Exley MA, Swain PA, and Briner TJ (1996). Efficacy of a therapeutic cocaine vaccine in rodent models. *Nat Med* 2(10): 1129-1132.

Franks P, Harp J, and Bell B (1989). Randomized, controlled trial of clonidine for smoking cessation in a primary care setting. *JAMA* 262(21): 3011-3013.

Frewer LJ (1990). The effect of betel nut on human performance. *P N G Med J* 33(2): 143-145.

Fukumoto M, Kubo H, and Ogamo A (1997). Determination of nicotine content of popular cigarettes. *Vet Hum Toxicol* 39(4): 225-227.

Garfinkel BD, Webster CD, and Sloman L (1981). Responses to methylphenidate and varied doses of caffeine in children with attention deficit disorder. *Can J Psychiatry* 26(6): 395-401.

Ghatan PH, Ingvar M, Eriksson L, Stone-Elander S, Serrander M, Ekberg K, and Wahren J (1998). Cerebral effects of nicotine during cognition in smokers and non-smokers. *Psychopharmacology (Berl)* 136(2): 179-189.

Gilbert DG, Robinson JH, Chamberlin CL, and Spielberger CD (1989). Effects of smoking/nicotine on anxiety, heart rate, and lateralization of EEG during a stressful movie. *Psychophysiology* 26(3): 311-320.

Gilbert RM, Marshman JA, Schwieder M, and Berg R (1976). Caffeine content of beverages as consumed. *Can Med Assoc J* 114(3): 205-208.

Glennon RA, Young R, Martin BR, and Dal Cason TA (1995). Methcathione ("cat"): An enantiomeric potency comparison. *Pharmacol Biochem Behav* 50(4): 601-606.

Golding J (1995). Reproduction and caffeine consumption—a literature review. *Early Hum Dev* 43(1): 1-14.

Grassi G, Seravalle G, Calhoun DA, Bolla GB, Giannattasio C, Marabini M, Del Bo A, and Mancia G (1994). Mechanisms responsible for sympathetic activation by cigarette smoking in humans. *Circulation* 90(1): 248-253.

Grobbee DE, Rimm EB, Giovannucci E, Colditz G, Stampfer M, and Willett W (1990). Coffee, caffeine, and cardiovascular disease in men. *N Engl J Med* 323(15): 1026-1032.

Gruber AJ and Pope HG Jr. (1998). Ephedrine abuse among 36 female weightlifters. *Am J Addict* 7(4): 256-261.

Gruenwald J, Brendler T, and Jaenicke C (1998). *PDR for herbal medicines,* First edition. Montvale, NJ: Medical Economics Company.

Hatsukami DK and Fischman MW (1996). Crack cocaine and cocaine hydrochloride. Are the differences myth or reality? *JAMA* 276(19): 1580-1588.

Hoff AL, Riordan H, Morris L, Cestaro V, Wieneke M, Alpert R, Wang GJ, and Volkow N (1996). Effects of crack cocaine on neurocognitive function. *Psychiatry Res* 60(2-3): 167-176.

Homstedt B, Lindgren JE, Rivier L, and Plowman T (1979). Cocaine in blood of coca chewers. *J Ethnopharmacol* 1(1): 69-78.

Hughes JR, Gust SW, Skoog K, Keenan RM, and Fenwick JW (1991). Symptoms of tobacco withdrawal: A replication and extension. *Arch Gen Psychiatry* 48(1): 52-59.

James JE (1998). Acute and chronic effects of caffeine on performance, mood, headache, and sleep. *Neuropsychobiology* 38(1): 32-41.

Jorenby DE, Hatsukami DK, Smith SS, Fiore MC, Allen S, Jensen J, and Baker TB (1996). Characterization of tobacco withdrawal symptoms: Transdermal nicotine reduces hunger and weight gain. *Psychopharmacology (Berl)* 128(2): 130-138.

Kalix P (1991). The pharmacology of psychoactive alkaloids from ephedra and catha. *J Ethnopharmacol* 32(1-3): 201-208.

Kalix P (1992). Cathinone, a natural amphetamine. *Pharmacol Toxicol* 70(2): 77-86.

Kaplan GB, Greenblatt DJ, Ehrenberg BL, Goddard JE, Cotreau MM, Harmatz JS, and Shader RI (1997). Dose-dependent pharmacokinetics and psychomotor effects of caffeine in humans. *J Clin Pharmacol* 37(8): 693-703.

Kennedy JG, Teague J, Rokaw W, and Cooney E (1983). A medical evaluation of the use of qat in North Yemen. *Soc Sci Med* 17(12): 783-793.

Knott VJ, Harr A, Ilivitsky V, and Mahoney C (1998). The cholinergic basis of the smoking-induced EEG activation profile. *Neuropsychobiology* 38(2): 97-107.

Leech SL, Richardson GA, Goldschmidt L, and Day NL (1999). Prenatal substance exposure: Effects on attention and impulsivity of 6-year-olds. *Neurotoxicol Teratol* 21(2): 109-118.

Leshner AI and Koob GF (1999). Drugs of abuse and the brain. *Proc Assoc Am Physicians* 111(2): 99-108.

Lieberman HR, Wurtman RJ, Emde GG, and Coviella IL (1987). The effects of caffeine and aspirin on mood and performance. *J Clin Psychopharmacol* 7(5): 315-320.

Linde L (1995). Mental effects of caffeine in fatigued and non-fatigued female and male subjects. *Ergonomics* 38(5): 864-885.

Mathew RJ, Barr DL, and Weinman ML (1983). Caffeine and cerebral blood flow. *Br J Psychiatry* 143: 604-608.

Nanji AA and Filipenko JD (1984). Asystole and ventricular fibrillation associated with cocaine intoxication. *Chest* 85(1): 132-133.

Nelson CB and Wittchen HU (1998). Smoking and nicotine dependence: Results from a sample of 14- to 24-year-olds in Germany. *Eur Addict Res* 4(1-2): 42-49.

Nordberg A, Lundqvist H, Hartvig P, Andersson J, Johansson M, Hellstrom-Lindahi E, and Norton SA (1998). Betel: Consumption and consequences. *J Am Acad Dermatol* 38(1): 81-88.

O'Brien CP (1996). Drug addiction and abuse. In Hardman JG, Limbird LE, Molinoff PB, Ruddon RW, and Goodman Gilman A (Eds.), *Goodman and Gilman's the pharmacological basis of therapeutics* (pp. 557-578). New York: McGraw-Hill.

O'Malley S, Adamse M, Heaton RK, and Gawin FH (1992). Neuropsychological impairment in chronic cocaine abusers. *Am J Drug Alcohol Abuse* 18(2): 131-144.

Pickworth WB, Fant RV, Butschky MF, and Henningfield JE (1996). Effects of transdermal nicotine delivery on measures of acute nicotine withdrawal. *J Pharmacol Exp Ther* 279(2): 450-456.

Pineda JA, Herrera C, Kang C, and Sandler A (1998). Effects of cigarette smoking and 12-h abstention on working memory during a serial-probe recognition task. *Psychopharmacology (Berl)* 139(4): 311-321.

Pirich C, O'Grady J, and Sinzinger H (1993). Coffee, lipoproteins and cardiovascular disease. *Wien Klin Wochenschr* 105(1): 3-6.

Pomerleau OF, Fertig JB, Seyler LE, and Jaffe J (1983). Neuroendocrine reactivity to nicotine in smokers. *Psychopharmacology (Berl)* 81(1): 61-67.

Rapoport JL, Berg CJ, Ismond DR, Zahn TP, and Neims A (1984). Behavioral effects of caffeine in children: Relationship between dietary choice and effects of caffeine challenge. *Arch Gen Psychiatry* 41(11): 1073-1079.

Reavill C, Walther B, Stolerman IP, and Testa B (1990). Behavioural and pharmacokinetic studies on nicotine, cytisine and lobeline. *Neuropharmacology* 29(7): 619-624.

Rudgley R (1999). *The encyclopaedia of psychoactive substances.* New York: St. Martin's Press.

Sachs DP and Leischow SJ (1991). Pharmacologic approaches to smoking cessation. *Clin Chest Med* 12(4): 769-791.

Schellscheidt J, Oyen N, and Jorch G (1997). Interactions between maternal smoking and other prenatal risk factors for sudden infant death syndrome (SIDS). *Acta Paediatr* 86(8): 857-863.

Schlyer D, Wolf AP, Warner D, Zezulkova I, and Cilento R (1996). Inhibition of monoamine oxidase B in the brains of smokers. *Nature* 379(6567): 733-736.

Sekkadde Kiyingi K and Saweri A (1994). Betelnut chewing causes bronchoconstriction in some asthma patients. *P N G Med J* 37(2): 90-99.

Selby MJ and Azrin RL (1998). Neuropsychological functioning in drug abusers. *Drug Alcohol Depend* 50(1): 39-45.

Shapira B, Lerer B, Gilboa D, Drexler H, Kugelmass S, and Calev A (1987). Facilitation of ECT by caffeine pretreatment. *Am J Psychiatry* 144(9): 1199-1202.

Shear MK (1986). Pathophysiology of panic: A review of pharmacologic provocative tests and naturalistic monitoring data. *J Clin Psychiatry* 47(Suppl.): 18-26.

Shively CA and Tarka SM Jr. (1984). Methylxanthine composition and consumption patterns of cocoa and chocolate products. *Prog Clin Biol Res* 158: 149-178.

Smith A, Kendrick A, Maben A, and Salmon J (1994). Effects of breakfast and caffeine on cognitive performance, mood and cardiovascular functioning. *Appetite* 22(1): 39-55.

Snyder SH and Sklar P (1984). Behavioral and molecular actions of caffeine: Focus on adenosine. *J Psychiatr Res* 18(2): 91-106.

Spindel ER, Wurtman RJ, McCall A, Carr DB, Conlay L, Griffith L, and Arnold MA (1984). Neuroendocrine effects of caffeine in normal subjects. *Clin Pharmacol Ther* 36(3): 402-407.

Stolerman IP and Shoaib M (1991). The neurobiology of tobacco addiction. *Trends Pharmacol Sci* 12(12): 467-473.

Terry WS and Phifer B (1986). Caffeine and memory performance on the AVLT. *J Clin Psychol* 42(6): 860-863.

Tyler VE (1994). *Herbs of choice: The therapeutic use of phytomedicinals.* Binghamton, NY: The Haworth Press.

Vidal C (1996). Nicotinic receptors in the brain: Molecular biology, function, and therapeutics. *Mol Chem Neuropathol* 28(1-3): 3-11.

Volkow ND, Fowler JS, and Wolf AP (1991). Use of positron emission tomography to study cocaine in the human brain. *NIDA Res Monogr* 112:168-179.

White LM, Gardner SF, Gurley BJ, Marx MA, Wang PL, and Estes M (1997). Pharmacokinetics and cardiovascular effects of ma-huang *(Ephedra sinica)* in normotensive adults. *J Clin Pharmacol* 37(2): 116-122.

Woolverton WL and Johnson KM (1992). Neurobiology of cocaine abuse. *Trends Pharmacol Sci* 13(5): 193-200.

Chapter 5

Adams RD, Victor M, and Ropper AH (1997). *Principles of neurology.* New York: McGraw-Hill.

Amri H, Drieu K, and Papadopoulos V (1997). Ex vivo regulation of adrenal cortical cell steroid and protein synthesis, in response to adrenocorticotropic hormone stimulation, by the *Ginkgo biloba* extract EGb 761 and isolated ginkgolide B. *Endocrinology* 138(12): 5415-5426.

Arendash GW, Sanberg PR, and Sengstock GJ (1995). Nicotine enhances the learning and memory of aged rats. *Pharmacol Biochem Behav* 52(3): 517-523.

Attele AS, Wu JA, and Yuan CS (1999). Ginseng pharmacology: Multiple constituents and multiple actions. *Biochem Pharmacol* 58(11): 1685-1693.

Babikian VL, Wolfe N, Lin R, Knoepal JE, and Albert ML (1990). Cognitive changes in patients with multiple cerebral infarcts. *Stroke* 21: 1013-1018.

Benishin CG (1992). Actions of ginsenoside Rb1 on choline uptake in central cholinergic nerve endings. *Neurochem Int* 21(1): 1-5.

Benishin CG, Lee R, Wang LC, and Liu HJ (1991). Effects of ginsenoside Rb1 on central cholinergic metabolism. *Pharmacology* 42(4): 223-229.

Biber A and Koch E (1999). Bioavailability of ginkgolides and bilobalide from extracts of *Ginkgo biloba* using GC/MS. *Planta Med* 65(2): 192-193.

Boller F and Duyckaerts C (1997). Alzheimer's disease: Clinical and anatomic aspects. In Feinberg T and Farah MJ (Eds.), *Behavioral neurology and neuropsychology* (pp. 1583-1622). New York: McGraw-Hill.

Bourgain RH, Maes L, Andries R, and Braquet P (1986). Thrombus induction by endogenic paf-acether and its inhibition by *Ginkgo biloba* extracts in the guinea pig. *Prostaglandins* 32(1): 142-144.

Bowler JV and Hachinski V (1997). Vascular dementia. In Feinberg T and Farah MJ (Eds.), *Behavioral neurology and neuropsychology* (pp. 589-603). New York: McGraw-Hill.

Brunello N, Racagni G, Clostre F, Drieu K, and Braquet P (1985). Effects of an extract of *Ginkgo biloba* on noradrenergic systems of rat cerebral cortex. *Pharmacol Res Commun* 17(11): 1063-1072.

Burke DM and Mackay DG (1997). Memory, language, and ageing. *Trans R Philos Soc Lond B Biol Sci* 352(1363): 1845-1856.

Cesarani A, Meloni F, Alpini D, Barozzi S, Verderio L, and Boscani PF (1998). *Ginkgo biloba* (EGb 761) in the treatment of equilibrium disorders. *Adv Ther* 15(5): 291-304.

Chen SE, Sawchuk RJ, and Staba EJ (1980). American ginseng. III. Pharmacokinetics of ginsenosides in the rabbit. *Eur J Drug Metab Pharmacokinet* 5(3): 161-168.

Choi YD, Xin ZC, and Choi HK (1998). Effect of Korean red ginseng on the rabbit corpus cavernosal smooth muscle. *Int J Impot Res* 10(1): 37-43.

Cohen AJ and Bartlik B (1998). *Ginkgo biloba* for antidepressant-induced sexual dysfunction. *J Sex Marital Ther* 124(2): 139-143.

Cohen-Salmon C, Venault P, Martin B, Raffalli-Sebille MJ, Barkats M, Clostre F, Pardon MC, Christen Y, and Chapouthier G (1997). Effects of *Ginkgo biloba* extract (EGb 761) on learning and possible actions on aging. *J Physiol Paris* 91(6): 291-300.

de Kloet ER, Oitzl MS, and Joels M (1999). Stress and cognition: Are corticosteroids good or bad guys? *Trends Neurosci* 22(10): 422-426.

Duche JC, Barre J, Guinot P, Duchier J, Cournot A, and Tillement JP (1989). Effect of *Ginkgo biloba* extract on microsomal enzyme induction. *Int J Clin Pharmacol Res* 9(3): 165-168.

Fahim MS, Fahim Z, Harman JM, Clevenger TE, Mullins W, and Hafez ES (1982). Effect of *Panax ginseng* on testosterone level and prostate in male rats. *Arch Androl* 8(4): 261-263.

Field B and Vadnal R (1998). *Ginkgo biloba* and memory: An overview. *Nutr Neurosci* 1: 2565-2567.

Foulds J, Stapleton J, Swettenham J, Bell N, McSorley K, and Russell MA (1996). Cognitive performance effects of subcutaneous nicotine in smokers and never-smokers. *Psychopharmacology (Berl)* 127(1): 31-38.

Fourtillan JB, Brisson AM, Girault J, Ingrand I, Decourt JP, Drieu K, Jouenne P, and Biber A (1995). [Pharmacokinetic properties of bilobalide and ginkgolides A and B in healthy subjects after intravenous and oral administration of *Ginkgo biloba* extract (EGb 761)]. *Therapie* 50(2): 137-144.

Gajewski A and Hensch SA (1999). *Ginkgo biloba* and memory for a maze. *Psychol Rep* 84(2): 481-484.

Gessner B, Voelp A, and Klasser M (1985). Study of the long-term action of a *Ginkgo biloba* extract on vigilance and mental performance as determined by means of quantitative pharmaco-EEG and psychometric measurements. *Arzneimittelforschung* 35(9): 1459-1465.

Gibb WRG (1998). The neuropathology of Parkinsonian disorders. In Jankovic J and Tolosa E (Eds.), *Parkinson's disease and movement disorders,* Third edition (pp. 253-270). Baltimore: Williams and Wilkins.

Gillis CN (1997). *Panax ginseng* pharmacology: A nitric oxide link? *Biochem Pharmacol* 54(1): 1-8.

Gross TM, Jarvik ME, and Rosenblatt MR (1993). Nicotine abstinence produces content-specific Stroop interference. *Psychopharmacology (Berl)* 110(3): 333-336.

Gruenwald J, Brendler T, and Jaenicke C (1998). *PDR for herbal medicines,* First edition. Montvale, NJ: Medical Economics Company.

Hiai S, Yokoyama H, Oura H, and Yano S (1979). Stimulation of pituitary-adrenocortical system by ginseng saponin. *Endocrinol Jpn* 26(6): 661-665.

Hindmarch I (1986). [Activity of *Ginkgo biloba* extract on short-term memory]. *Presse Medicale* 15(31): 1592-1594.

Itil T and Martorano D (1995). Natural substances in psychiatry (*Ginkgo biloba* in dementia). *Psychopharmacol Bull* 31(1): 147-158.

Itil TM, Eralp E, Ahmed I, Kunitz A, and Itil KZ (1998). The pharmacological effects of *Ginkgo biloba,* a plant extract, on the brain of dementia patients in comparison with tacrine. *Psychopharmacol Bull* 34(3): 391-397.

Itoh T, Zang YF, Murai S, and Saito H (1989). Effects of *Panax ginseng* root on the vertical and horizontal motor activities and on brain monoamine-related substances in mice. *Planta Med* 55(5): 429-433.

Jagust WJ, Reed BR, Martin EM, Eberling JL, and Nelson-Abbott RA (1992). Cognitive function and regional cerebral blood flow in Parkinson's disease. *Brain* 115(Pt 2): 521-537.

Joyce CA, Paller KA, McIsaac HK, and Kutas M (1998). Memory changes with normal aging: Behavioral and electrophysiological measures. *Psychophysiology* 35(6): 669-678.

Jung F, Mrowietz C, Kiesewetter H, and Wenzel E (1990). Effect of *Ginkgo biloba* on fluidity of blood and peripheral microcirculation in volunteers. *Arzneimittelforschung* 40(5): 589-593.

Kanowski S, Herrmann WM, Stephan K, Wierich W, and Hörr R (1996). Proof of efficacy of the *Ginkgo biloba* special extract EGb 761 in outpatients suffering from mild to moderate primary degenerative dementia of the Alzheimer type or multi-infarct dementia. *Pharmacopsychiatry* 29(2): 47-56.

Karplus TM and Saag KG (1998). Nonsteroidal anti-inflammatory drugs and cognitive function: Do they have a beneficial or deleterious effect? *Drug Saf* 19(6): 427-433.

Keefover RW (1998). Aging and cognition. *Neurol Clin* 16(3): 635-648.

Kleijnen J and Knipschild P (1992). *Ginkgo biloba. Lancet* 340(8828): 1136-1139.

Konno S, Meyer JS, Terayama Y, Margishvili GM, and Mortel KF (1997). Classification, diagnosis and treatment of vascular dementia. *Drugs Aging* 11(5): 361-373.

Krall WJ, Sramek JJ, and Cutler NR (1999). Cholinesterase inhibitors: A therapeutic strategy for Alzheimer disease. *Ann Pharmacother* 33(4): 441-450.

Kristofikova Z, Benesova O, and Tejkalova H (1992). Changes in high-affinity choline uptake in the hippocampus of old rats after long-term administration of two nootropic drugs (tacrine and *Ginkgo biloba* extract). *Dementia* 3: 304-307.

Kristofikova Z and Klaschka J (1997). In vitro effect of *Ginkgo biloba* extract (EGb 761) on the activity of presynaptic cholinergic nerve terminals in rat hippocampus. *Dementia Geriatr Cogn Disord* 8(1): 43-48.

Kunkel H (1993). EEG profile of three different extractions of *Ginkgo biloba*. *Neuropsychobiology* 27(1): 40-45.

Lawrence AD and Sahakian BJ (1998). The cognitive psychopharmacology of Alzheimer's disease: Focus on cholinergic systems. *Neurochem Res* 23(5): 787-794.

Le Houezec J (1998). Nicotine: Abused substance and therapeutic agent. *J Psychiatry Neurosci* 23(2): 95-108.

Li CL and Wong YY (1997). The bioavailability of ginkgolides in *Ginkgo biloba* extracts. *Planta Med* 63(6): 563-565.

Lugasi A, Horvahovich P, and Dworschak E (1999). Additional information to the in vitro antioxidant activity of *Ginkgo biloba* L. *Phytother Res* 13(2): 160-162.

Maitra I, Marcocci L, Droy-Lefaix MT, and Packer L (1995). Peroxyl radical scavenging activity of *Ginkgo biloba* extract EGb 761. *Biochem Pharmacol* 49(11): 1649-1655.

Manly JJ, Jacobs DM, Sano M, Bell K, Merchant CA, Small SA, and Stern Y (1999). Effect of literacy on neuropsychological test performance in nondemented, education-matched elders. *J Int Neuropsychol Soc* 5(3): 191-202.

Marcilhac A, Dakine N, Bourhim N, Guillaume V, Grino M, Drieu K, and Oliver C (1998). Effect of chronic administration of *Ginkgo biloba* extract or ginkgolide on the hypothalamic-pituitary-adrenal axis in the rat. *Life Sci* 62(25): 2329-2340.

Nyenhuis DL and Gorelick PB (1998). Vascular dementia: A contemporary review of epidemiology, diagnosis, prevention, and treatment. *J Am Geriatr Soc* 46(11): 1437-1448.

Oken BS, Storzbach DM, and Kaye JA (1998). The efficacy of *Ginkgo biloba* on cognitive function in Alzheimer disease. *Arch Neurol* 55(11): 1409-1415.

Olsen CG and Clasen ME (1998). Senile dementia of the Binswanger's type. *Am Fam Physician* 58(9): 2068-2074.

Oyama Y, Chikahisa L, Ueha T, Kanemaru K, and Noda K (1996). *Ginkgo biloba* extract protects brain neurons against oxidative stress induced by hydrogen peroxide. *Brain Res* 712(2): 349-352.

Paick JS and Lee JH (1996). An experimental study of the effect of *Ginkgo biloba* extract on the human and rabbit corpus cavernosum tissue. *J Urol* 156: 1876-1880.

Petkov VD, Cao Y, Todorov I, Lazarova M, Getova D, Stancheva S, and Alova L (1992). Behavioral effects of stem-leaves extract from *Panax ginseng* C.A. Meyer. *Acta Physiol Pharmacol Bulg* 18(2): 41-48.

Petkov VD, Kehayov R, Belcheva S, Konstantinova E, Petkov VV, Getova D, and Markovska V (1993). Memory effects of standardized extracts of *Panax ginseng* (G115), *Ginkgo biloba* (GK 501) and their combination Gincosan (PHL-00701). *Planta Med* 59(2): 106-114.

Petkov VD and Mosharrof AH (1987). Effects of standardized ginseng extract on learning, memory and physical capabilities. *Am J Chin Med* 15(1-2): 19-29.

Pidoux B (1986). [Effects of *Ginkgo biloba* extract on functional brain activity: An assessment of clinical and experimental studies]. *Presse Med* 15(31): 1588-1591.

Pietta PG, Gardana C, and Mauri PL (1997). Identification of *Gingko biloba* flavonol metabolites after oral administration to humans. *J Chromatogr B Biomed Sci Appl* 693(1): 249-255.

Rabey JM, Vered Y, Shabtai H, Graff F, Harsat A, and Korczyn AD (1993). Broad bean *(Vicia faba)* consumption and Parkinson's disease. *Ad Neurol* 60: 681-684.

Rabey JM, Vered Y, Shabtai H, Graff F, and Korczyn AD (1992). Improvement of parkinsonian features correlate with high plasma levodopa values after broad bean *(Vicia faba)* consumption. *J Neurol Neurosurg Psychiatry* 55(8): 725-727.

Ramassamy C, Christen Y, Clostre F, and Costentin J (1992). The *Ginkgo biloba* extract, EGb761, increases synaptosomal uptake of 5-hydroxytryptamine: In-vitro and ex-vivo studies. *J Pharm Pharmacol* 44(11): 943-945.

Rapin JR, Lamproglou I, Drieu K, and DeFeudis FV (1994). Demonstration of the "anti-stress" activity of an extract of *Ginkgo biloba* (EGb 761) using a discrimination learning task. *Gen Pharmacol* 25(5): 1009-1016.

Rich JB, Rasmusson DX, Folstein MF, Carson KA, Kawas C, and Brandt J (1995). Nonsteroidal anti-inflammatory drugs in Alzheimer's disease. *Neurology* 45(1): 51-55.

Robbers JE, Speedie MK, and Tyler VE (1996). *Pharmacognosy and pharmacobiotechnology*. Baltimore: Williams and Wilkins.

Robbins TW, McAlonan G, Muir JL, and Everitt BJ (1997). Cognitive enhancers in theory and practice: Studies of the cholinergic hypothesis of cognitive deficits in Alzheimer's disease. *Behav Brain Res* 83(1-2): 15-23.

Salim KN, McEwen BS, and Chao HM (1997). Ginsenoside Rb1 regulates ChAT, NGF and trkA mRNA expression in the rat brain. *Brain Res Mol Brain Res* 47 (1-2): 177-182.

Sano M, Ernesto C, Thomas RG, Klauber MR, Schafer K, Grundman M, Woodbury P, Growdon J, Cotman CW, Pfeiffer E, Schneider LS, and Thal LJ (1997). A controlled trial of selegiline, alpha-tocopherol, or both as treatment for Alzheimer's disease: The Alzheimer's Disease Cooperative Study. *N Engl J Med* 336(17): 1216-1222.

Siegel RK (1979). Ginseng abuse syndrome. Problems with the panacea. *JAMA* 241(15): 1614-1615.

Snyder FR and Henningfield JE (1989). Effects of nicotine administration following 12 h of tobacco deprivation: Assessment on computerized performance tasks. *Psychopharmacology (Berl)* 97(1): 17-22.

Soholm B (1998). Clinical improvement of memory and other cognitive functions by *Ginkgo biloba:* Review of relevant literature. *Adv Ther* 15(1): 54-65.

Spengos M and Vassilopoulos D (1993). Improvement of Parkinson's disease after *Vicia faba* consumption. Ninth International Symposium on Parkinson's disease. *Asia Pacific J Clin Nutr* 2:85-89.

Standaert DG and Young AB (1996). Treatment of central nervous system degenerative disorders. In Hardman JG, Limbird LE, Molinoff PB, Ruddon RW, and Goodman Gilman A (Eds.), *Goodman and Gilman's the pharmacological basis of therapeutics* (pp. 506-513). New York: McGraw-Hill.

Stewart WF, Kawas C, Corrada M, and Metter EJ (1997). Risk of Alzheimer's disease and duration of NSAID use. *Neurology* 48(3): 626-632.

Subhan Z and Hindmarch I (1984). The psychopharmacological effects of *Ginkgo biloba* extract in normal healthy volunteers. *Int J Clin Pharmacol Res* 4(2): 89-93.

Taylor AE, Saint-Cyr JA, and Lang AE (1986). Frontal lobe dysfunction in Parkinson's disease: The cortical focus of neostriatal outflow. *Brain* 109(Pt. 5): 845-883.

Warburton DM (1992). Nicotine as a cognitive enhancer. *Prog Neuropsychopharmacol Biol Psychiatry* 16(2): 181-191.

White HL, Scates PW, and Cooper BR (1996). Extracts of *Ginkgo biloba* leaves inhibit monoamine oxidase. *Life Sci* 58(16): 1315-1321.

Winther K, Randlov C, Rein E, and Mehlsen J (1998). Effects of *Ginkgo biloba* extract on cognitive function and blood pressure in elderly subjects. *Curr Ther Res* 59(12): 881-888.

Wójcicki J, Gawroska-Szklarz B, Bieganowski W, Patalan M, Smulski HK, Samochowiec L, and Zakrzewski J (1995). Comparative pharmacokinetics and bioavailability of flavonoid glycosides of *Ginkgo biloba* after a single oral administration of three formulations to healthy volunteers. *Mater Med Pol* 27(4): 141-146

Zhao R and McDaniel WF (1998). Ginseng improves strategic learning by normal and brain-damaged rats. *Neuroreport* 9(7): 1619-1624.

Chapter 6

Backhauss C and Krieglstein J (1992). Extract of kava *(Piper methysticum)* and its methysticin constituents protect brain tissue against ischemic damage in rodents. *Eur J Pharmacol* 215(2-3): 265-269.

Balderer G and Borbely AA (1985). Effect of valerian on human sleep. *Psychopharmacology (Berl)* 87(4): 406-409.

Baum SS, Hill R, and Rommelspacher H (1998). Effect of kava extract and individual kavalactones on neurotransmitter levels in the nucleus accumbens of rats. *Prog Neuropsychopharmacol Biol Psychiatry* 22(7): 1105-1120.

Boonen G, Ferger B, Kuschinsky K, and Haberlein H (1998). In vivo effects of the kavalactones (+)-dihydromethysticin and (+/-)-kavain on dopamine, 3,4-dihydroxyphenylacetic acid, serotonin and 5-hydroxyindoleacetic acid levels in striatal and cortical brain regions. *Planta Med* 64(6): 507-510.

Boonen G and Haberlein H (1998). Influence of genuine kavalactone enantiomers on the GABA-A binding site. *Planta Med* 64(6): 504-506.

Bourdet C and Goldenberg F (1994). Insomnia in anxiety: Sleep EEG changes. *J Psychosom Res* 38(Suppl. 1): 93-104.

Capasso A, Piacente S, Pizza C, and Sorrentino L (1998). Flavonoids reduce morphine withdrawal in-vitro. *J Pharm Pharmacol* 50(5): 561-564.

Davies LP, Drew CA, Duffield P, Johnston GA, and Jamieson DD (1992). Kava pyrones and resin: Studies on GABAA, GABAB and benzodiazepine binding sites in rodent brain. *Pharmacol Toxicol* 71(2): 120-126.

Foo H and Lemon J (1997). Acute effects of kava, alone or in combination with alcohol, on subjective measures of impairment and intoxication and on cognitive performance. *Drug Alcohol Rev* 16(2): 147-155.

Frey R (1991). [Demonstration of the central effects of D,L-kawain with EEG brain mapping]. *Fortschr Med* 109(25): 505-508.

Friese J and Gleitz J (1998). Kavain, dihydrokavain, and dihydromethysticin noncompetitively inhibit the specific binding of [3H]-batrachotoxinin-A 20-alpha-benzoate to receptor site 2 of voltage-gated Na+ channels. *Planta Med* 64(5): 458-459.

Gleitz J, Beile A, and Peters T (1995). (+/-)-Kavain inhibits veratridine-activated voltage-dependent Na(+)-channels in synaptosomes prepared from rat cerebral cortex. *Neuropharmacology* 34(9): 1133-1138.

Gleitz J, Friese J, Beile A, Ameri A, and Peters T (1996). Anticonvulsive action of (+/-)-kavain estimated from its properties on stimulated synaptosomes and Na+ channel receptor sites. *Eur J Pharmacol* 315(1): 89-97.

Gruenwald J, Brendler T, and Jaenicke C (1998). *PDR for herbal medicines*, First edition. Montvale, NJ: Medical Economics Company.

Hänsel R and Wagener HH (1967). [Attempts to identify sedative-hypnotic active substances in hops]. *Arzneimittelforschung* 17(1): 79-81.

Harney JW, Barofsky IM, and Leary JD (1978). Behavioral and toxicological studies of cyclopentanoid monoterpenes from *Nepeta cataria*. *Lloydia* 41(4): 367-374.

Hart BL and Leedy MG (1985). Analysis of the catnip reaction: Mediation by olfactory system, not vomeronasal organ. *Behav Neural Biol* 44(1): 38-46.

Heinze HJ, Münthe TF, Steitz J, and Matzke M (1994). Pharmacopsychological effects of oxazepam and kava-extract in a visual search paradigm assessed with event-related potentials. *Pharmacopsychiatry* 27(6): 224-230.

Hobbs WR, Rall TW, and Verdoon TA (1996). Hypnotics and sedatives: Ethanol. In Hardman JG, Limbird LE, Molinoff PB, Ruddon RW, and Goodman Gilman A (Eds.), *Goodman and Gilman's the pharmacololgical basis of therapeutics* (pp. 361-398). New York: McGraw-Hill.

Holm E, Staedt U, Heep J, Kortsik C, Behne F, Kaske A, and Mennicke I (1991). [The action profile of D,L-kavain. Cerebral sites and sleep-wakefulness rhythm in animals]. *Arzneimittelforschung* 41(7): 673-683.

Houghton PJ (1988). The biological activity of valerian and related plants. *J Ethnopharmacol* 22(2): 121-142.

Houghton PJ (1999). The scientific basis for the reputed activity of valerian. *J Pharm Pharmacol* 51(5): 505-512.

Jamieson DD and Duffield PH (1990). Positive interaction of ethanol and kava resin in mice. *Clin Exp Pharmacol Physiol* 17(7): 509-514.

Jamieson DD, Duffield PH, Cheng D, and Duffield AM (1989). Comparison of the central nervous system activity of the aqueous and lipid extract of kava *(Piper methysticum). Arch Int Pharmacodyn Ther* 301(September/October): 66-80.

Juhász G, Emri Z, Kékesi KA, Salfay O, and Crunelli V (1994). Blockade of thalamic GABAB receptors decreases EEG synchronization. *Neurosci Lett* 172(1-2): 155-158.

Jussofie A, Schmiz A, and Hiemke C (1994). Kavapyrone enriched extract from *Piper methysticum* as modulator of the GABA binding site in different regions of rat brain. *Psychopharmacology (Berl)* 116(4): 469-474.

Keledjian J, Duffield PH, Jamieson DD, Lidgard RO, and Duffield AM (1988). Uptake into mouse brain of four compounds present in the psychoactive beverage kava. *J Pharm Sci* 77(12): 1003-1006.

Leathwood PD and Chauffard F (1985). Aqueous extract of valerian reduces latency to fall asleep in man. *Planta Med* 2: 144-148.

Leathwood PD, Chauffard F, Heck E, and Munoz-Box R (1982). Aqueous extract of valerian root (*Valeriana officinalis* L.) improves sleep quality in man. *Pharmacol Biochem Behav* 17(1): 65-71.

Lebot V, Merlin M, and Lindstrom L ([1992] 1997). *Kava—the Pacific elixir: The definitive guide to its ethnobotany, history, and chemistry*. Rochester, VT: Healing Arts.

Lehmann E, Klieser E, Klimke A, Krach H, and Spatz R (1989). The efficacy of cavain in patients suffering from anxiety. *Pharmacopsychiatry* 22(6): 258-262.

Leuschner J, Müller J, and Rudmann M (1993). Characterisation of the central nervous depressant activity of a commercially available valerian root extract. *Arzneimittelforschung* 43(6): 638-641.

Lindahl O and Lindwall L (1989). Double blind study of a valerian preparation. *Pharmacol Biochem Behav* 32(4): 1065-1066.

Lindenberg D and Pitule-Schodel H (1990). [D,L-kavain in comparison with oxazepam in anxiety disorders. A double-blind study of clinical effectiveness]. *Fortschr Med* 108(2): 49-50, 53-54.

Magura EI, Kopanitsa MV, Gleitz J, Peters T, and Krishtal OA (1997). Kava extract ingredients, (+)-methysticin and (+/-)-kavain inhibit voltage-operated Na(+)-channels in rat CA1 hippocampal neurons. *Neuroscience* 81(2): 345-351.

Maluf E, Baros HMT, Frochtengarten ML, Benti R, and Leite JR (1991). Assessment of the hypnotic/sedative effects and toxicity of *Passiflora edulis* aqueous extract in rodents and humans. *Phytother Res* 5(6): 262-266.

Massoco CO, Silva MR, Gorniak SL, Spinosa MS, and Bernardi MM (1995). Behavioral effects of acute and long-term administration of catnip *(Nepeta cataria)* in mice. *Vet Hum Toxicol* 37(6): 530-533.

Medina JH, Paladini AC, Wolfman C, Levi de Stein M, Calvo D, Diaz LE, and Pena C (1990). Chrysin (5,7-di-OH-flavone), a naturally-occurring ligand for benzodiazepine receptors, with anticonvulsant properties. *Biochem Pharmacol* 40(10): 2227-2231.

Medina JH, Viola H, Wolfman C, Marder M, Wasowski C, Calvo D, and Paladini AC (1997). Overview—flavonoids: A new family of benzodiazepine receptor ligands. *Neurochem Res* 22(4): 419-425.

Meier B (1995). *Passiflora incarnata* L.—Passion flower. *Zeitschrift fur Phytotherapie* 16(2): 115-126.

Münte TF, Heinze HJ, Matzke M, and Steitz J (1993). Effects of oxazepam and an extract of kava roots *(Piper methysticum)* on event-related potentials in a word recognition task. *Neuropsychobiology* 27(1): 46-53.

Ortiz JG, Nieves-Natal J, and Chavez P (1999). Effects of *Valeriana officinalis* extracts on [3H]flunitrazepam binding, synaptosomal [3H]GABA uptake, and hippocampal [3H]GABA release. *Neurochem Res* 24(11):1373-1378.

Osterhoudt KC, Lee SK, Callahan JM, and Henretig FM (1997). Catnip and the alteration of human consciousness. *Vet Hum Toxicol* 39(6): 373-375.

Pittler MH and Ernst E (2000). Efficacy of kava extract for treating anxiety: Systematic review and meta-analysis. *J Clin Psychopharmacol* (1): 84-89.

Riedel E, Hansel R, and Ehrke G (1982). [Inhibition of gamma-aminobutyric acid catabolism by valerenic acid derivatives]. *Planta Med* 46(4): 219-220.

Salgueiro JB, Ardenghi P, Dias M, Ferreira MB, Izquierdo I, and Medina JH (1997). Anxiolytic natural and synthetic flavonoid ligands of the central benzodiazepine receptor have no effect on memory tasks in rats. *Pharmacol Biochem Behav* 58(4): 887-891.

Santos MS, Ferreira F, Cunha AP, Carvalho AP, and Macedo T (1994). An aqueous extract of valerian influences the transport of GABA in synaptosomes. *Planta Med* 60(3): 278-279.

Santos MS, Ferreira F, Cunha AP, Carvalho AP, Ribeiro CF, and Macedo T (1994). Synaptosomal GABA release as influenced by valerian root extract—Involvement of the GABA carrier. *Arch Int Pharmacodyn Ther* 327(2): 220-231.

Santos MS, Ferreira F, Faro C, Pires E, Carvalho AP, Cunha AP, and Macedo T (1994). The amount of GABA present in aqueous extracts of valerian is sufficient to account for [3H]GABA release in synaptosomes. *Planta Med* 60(5): 475-476.

Schelosky L, Raffauf C, Jendroska K, and Poewe W (1995). Kava and dopamine antagonism [letter]. *J Neurol Neurosurg Psychiatry* 58(5): 639-640.

Schirrmacher K, Busselberg D, Langosch JM, Walden J, Winter U, and Bingmann D (1999). Effects of (+/-)-kavain on voltage-activated inward currents of dorsal root ganglion cells from neonatal rats. *Eur Neuropsychopharmacol* 9(1-2): 171-176.

Schulz H, Stolz C, and Müller J (1994). The effect of valerian extract on sleep polygraphy in poor sleepers: A pilot study. *Pharmacopsychiatry* 27(4): 147-151.

Seitz U, Schule A, and Gleitz J (1997). [3H]-monoamine uptake inhibition properties of kava pyrones. *Planta Med* 63(6): 548-549.

Sherry CJ and Hunter PS (1979). The effect of an ethanol extract of catnip *(Nepeta cataria)* on the behavior of the young chick. *Experientia* 35(2): 237-238.

Sopranzi N, De Feo G, Mazzanti G, and Tolu L (1990). [Biological and electroencephalographic parameters in rats in relation to *Passiflora incarnata* L.]. *Clin Ther* 132(5): 329-333.

Temkin O (1971). *The falling sickness: A history of epilepsy from the Greeks to the beginnings of modern neurology*. Baltimore: Johns Hopkins University Press.

Tyler V (1994). *Herbs of choice*. Binghamton, NY: The Haworth Press.

Uebelhack R, Franke L, and Schewe HJ (1998). Inhibition of platelet MAO-B by kava pyrone-enriched extract from *Piper methysticum* Forster (kava-kava). *Pharmacopsychiatry* 31(5): 187-192.

Viola H, Wasowski C, Levi de Stein M, Wolfman C, Silveira R, Dajas F, Medina JH, and Paladini AC (1995). Apigenin, a component of *Matricaria recutita* flowers, is a central benzodiazepine receptors-ligand with anxiolytic effects. *Planta Med* 61(3): 213-216.

Wolfman C, Viola H, Paladini A, Dajas F, and Medina JH (1994). Possible anxiolytic effects of chrysin, a central benzodiazepine receptor ligand isolated from *Passiflora coerulea. Pharmacol Biochem Behav* 47(1): 1-4.

Wong AHC, Smith M, and Boon H (1998). Herbal remedies in psychiatric practice. *Arch Gen Psychiatry* 55(November): 1033-1044.

Chapter 7

American Psychiatric Association (1994). *Diagnostic and statistical manual of mental disorders,* Fourth edition. Washington, DC: American Psychiatric Association.

Amri H, Drieu K, and Papadopoulos V (1997). Ex vivo regulation of adrenal cortical cell steroid and protein synthesis, in response to adrenocorticotropic hormone stimulation, by the *Ginkgo biloba* extract EGb 761 and isolated ginkgolide B. *Endocrinology* 138(12): 5415-5426.

Amri H, Ogwuegbu SO, Boujrad N, Drieu K, and Papadopoulos V (1996). In vivo regulation of peripheral-type benzodiazepine receptor and glucocorticoid synthesis by *Ginkgo biloba* extract EGb 761 and isolated ginkgolides. *Endocrinology* 137(12): 5707-5718.

Arborelius L, Owens MJ, Plotsky PM, and Nemeroff CB (1999). The role of corticotropin-releasing factor in depression and anxiety disorders. *J Endocrinol* 160(1): 1-12.

Armario A, Marti O, Molina T, de Pablo J, and Valdes M (1996). Acute stress markers in humans: Response of plasma glucose, cortisol and prolactin to two examinations differing in the anxiety they provoke. *Psychoneuroendocrinology* 21(1): 17-24.

Artigas F, Romero L, de Montigny C, and Blier P (1996). Acceleration of the effect of selected antidepressant drugs in major depression by 5-HT_{1A} antagonists. *Trends Neurosci* 19(9): 378-383.

Bagdy G, Perenyi A, Frecska E, Seregi A, Fekete MI, Tothfalusi L, Magyar K, Bela A, and Arato M (1988). Effect of adjuvant reserpine treatment on catecholamine metabolism in schizophrenic patients under long-term neuroleptic treatment. *J Neural Transm* 71(1): 73-78.

Baldessarini R (1996). Drugs and the treatment of psychiatric disorders. In Hardman JG, Limbird LE, Molinoff PB, Ruddon RW, and Goodman Gilman A

(Eds.), *Goodman and Gilman's the pharmacological basis of therapeutics* (pp. 431-460). New York: McGraw-Hill.

Baureithel KH, Buter KB, Engesser A, Burkard W, and Schaffner W (1997). Inhibition of benzodiazepine binding in vitro by amentoflavone, a constituent of various species of hypericum. *Pharm Acta Helv* 72(3): 153-157.

Bhattacharya SK, Chakrabarti A, and Chatterjee SS (1998). Activity profiles of two hyperforin-containing hypericum extracts in behavioral models. *Pharmacopsychiatry* 31(Suppl. 1): 22-29.

Biber A, Fischer H, Romer A, and Chatterjee SS (1998). Oral bioavailability of hyperforin from hypericum extracts in rats and human volunteers. *Pharmacopsychiatry* 31(Suppl. 1): 36-43.

Bladt S and Wagner H (1994). Inhibition of MAO by fractions and constituents of hypericum extract. *J Geriatr Psychiatry Neurol* 7(Suppl. 1): S57-S59.

Blazer DG, Kessler RC, McGonagle KA, and Swartz MS (1994). The prevalence and distribution of major depression in a national community sample: The National Comorbidity Survey. *Am J Psychiatry* 151: 979-986.

Bloom FE and Morales M (1998). The central 5-HT3 receptor in CNS disorders. *Neurochem Res* 23(5): 653-659.

Bolanos-Jimenez F, Manhaes de Castro R, Sarhan H, Prudhomme N, Drieu K, and Fillion G (1995). Stress-induced 5-HT1A receptor desensitization: Protective effects of *Ginkgo biloba* extract (EGb 761). *Fundam Clin Pharmacol* 9(2): 169-174.

Brockmöller J, Reum T, Bauer S, Kerb R, Hübner WD, and Roots I (1997). Hypericin and pseudohypericin: Pharmacokinetics and effects on photosensitivity in humans. *Pharmacopsychiatry* 30(Suppl. 2): 94-101.

Butterweck V, Wall A, Lieflander-Wulf U, Winterhoff H, and Nahrstedt A (1997). Effects of the total extract and fractions of *Hypericum perforatum* in animal assays for antidepressant activity. *Pharmacopsychiatry* 30(Suppl. 2): 117-124.

Charney DS (1998). Monoamine dysfunction and the pathophysiology and treatment of depression. *J Clin Psychiatry* 59(Suppl. 14): 11-14.

Chatterjee SS, Bhattacharya SK, Wonnemann M, Singer A, and Muller WE (1998). Hyperforin as a possible antidepressant component of hypericum extracts. *Life Sci* 63(6): 499-510.

Chatterjee SS, Noldner M, Koch E, and Erdelmeier C (1998). Antidepressant activity of *Hypericum perforatum* and hyperforin: The neglected possibility. *Pharmacopsychiatry* 31(Suppl. 1): 7-15.

Costall B and Naylor RJ (1992). Neuropharmacology of emesis in relation to clinical response. *Br J Cancer* 19: S2-S7, discussion S7-S8.

Costall B, Naylor RJ, and Tyers MB (1990). The psychopharmacology of 5-HT3 receptors. *Pharmacol Ther* 47(2): 181-202.

Cott JM (1997). In vitro receptor binding and enzyme inhibition by *Hypericum perforatum* extract. *Pharmacopsychiatry* 30(Suppl. 2): 108-112.

Ding GH, Naora K, Hayashibara M, Katagiri Y, Kano Y, and Iwamoto K (1991). Pharmacokinetics of [6]-gingerol after intravenous administration in rats. *Chem Pharm Bull (Tokyo)* 39(6): 1612-1614.

Eison MS (1990). Serotonin: A common neurobiologic substrate in anxiety and depression. *J Clin Psychopharmacol* 10(3 Suppl.): 26S-30S.

Ernst E, Rand JI, Barnes J, and Stevinson C (1998). Adverse effects profile of the herbal antidepressant St. John's wort (*Hypericum perforatum* L.). *Eur J Clin Pharmacol* 54(8): 589-594.

Fowler JS, Wang GJ, Volkow ND, Logan J, Franceschi D, Franceschi M, MacGregor R, Shea C, Garza V, Liu N, and Ding YS (2000). Evidence that *Ginkgo biloba* extract does not inhibit MAO A and B in living human brain. *Life Sci* 66(9): PL141-PL146.

Gerlach J (1991). New antipsychotics: Classification, efficacy, and adverse effects. *Schizophr Bull* 17(2): 289-309.

Gordon JB (1998). SSRIs and St. John's wort: Possible toxicity? *Am Fam Physician* 57(5): 950-951.

Gruenwald J, Brendler T, and Jaenicke C (1998). *PDR for herbal medicines,* First edition. Montvale, NJ: Medical Economics Company.

Heninger GR, Delgado PL, and Charney DS (1996). The revised monoamine theory of depression: A modulatory role for monoamines, based on new findings from monoamine depletion experiments in humans. *Pharmacopsychiatry* 29(1): 2-11.

Johne A, Brockmoller J, Bauer S, Maurer A, Langheinrich M, and Roots I (1999). Pharmacokinetic interaction of digoxin with an herbal extract from St John's wort *(Hypericum perforatum). Clin Pharmacol Ther* 66(4): 338-345.

Kerb R, Brockmoller J, Staffeldt B, Ploch M, and Roots I (1996). Single-dose and steady-state pharmacokinetics of hypericin and pseudohypericin. *Antimicrob Agents Chemother* 40(9): 2087-2093.

Kim HL, Streltzer J, and Goebert D (1999). St. John's wort for depression: A meta-analysis of well-defined clinical trials. *J Nerv Ment Dis* 187(9): 532-538.

Laakmann G, Schule C, Baghai T, and Kieser M (1998). St. John's wort in mild to moderate depression: The relevance of hyperforin for the clinical efficacy. *Pharmacopsychiatry* 31(Suppl. 1): 54-59.

Linde K, Ramirez G, Mulrow CD, Pauls A, Weidenhammer W, and Melchart D (1996). St John's wort for depression—An overview and meta-analysis of randomised clinical trials. *BMJ* 313(7052): 253-258.

Lingjaerde O, Foreland AR, and Magnusson A (1999). Can winter depression be prevented by *Ginkgo biloba* extract? A placebo-controlled trial. *Acta Psychiatr Scand* 100(1): 62-66.

Marcilhac A, Dakine N, Bourhim N, Guillaume V, Grino M, Drieu K, and Oliver C (1998). Effect of chronic administration of *Ginkgo biloba* extract or ginkgolide on the hypothalamic-pituitary-adrenal axis in the rat. *Life Sci* 62(25): 2329-2340.

Muller WE, Rolli M, Schafer C, and Hafner U (1997). Effects of hypericum extract (LI 160) in biochemical models of antidepressant activity. *Pharmacopsychiatry* 30(Suppl. 2): 102-107.

Muller WE and Rossol R (1994). Effects of hypericum extract on the expression of serotonin receptors. *J Geriatr Psychiatry Neurol* 7(Suppl. 1): S63-S64.

Muller WE, Singer A, Wonnemann M, Hafner U, Rolli M, and Schafer C (1998). Hyperforin represents the neurotransmitter reuptake inhibiting constituent of hypericum extract. *Pharmacopsychiatry* 31(Suppl. 1): 16-21.

Mustafa T, Srivastava KC, and Jensen KB (1993). Pharmacology of ginger, *Zingiber officinale*. *J Drug Dev* 6(1): 25-39.

Naora K, Ding G, Hayashibara M, Katagiri Y, Kano Y, and Iwamoto K (1992). Pharmacokinetics of [6]-gingerol after intravenous administration in rats with acute renal or hepatic failure. *Chem Pharm Bull (Tokyo)* 40(5): 1295-1298.

Nebel A, Schneider BJ, Baker RK, and Kroll DJ (1999). Potential metabolic interaction between St. John's wort and theophylline. *Ann Pharmacother* 33(4): 502.

O'Breasail AM and Argouarch S (1998). Hypomania and St. John's wort. *Can J Psychiatry* 43(7): 746-747.

Porsolt RD, Martin P, Lenegre A, Fromage S, and Drieu K (1990). Effects of an extract of *Ginkgo biloba* (EGB 761) on "learned helplessness" and other models of stress in rodents. *Pharmacol Biochem Behav* 36(4): 963-971.

Robbers JE, Speedie MK, and Tyler VE (1996). *Pharmacognosy and pharmacobiotechnology*. Baltimore: Willams and Wilkins.

Rocha L, Marston A, Kaplan MA, Stoeckli-Evans H, Thull U, Testa B, and Hostettmann K (1994). An antifungal gamma-pyrone and xanthones with monoamine oxidase inhibitory activity from *Hypericum brasiliense*. *Phytochemistry* 36(6): 1381-1385.

Ruschitzka F, Meier PJ, Turina M, Luscher TF, and Noll G (2000). Acute heart transplant rejection due to Saint John's wort. *Lancet* 355(9203): 548-549.

Satyan KS, Jaiswal AK, Ghosal S, and Bhattacharya SK (1998). Anxiolytic activity of ginkgolic acid conjugates from Indian *Ginkgo biloba*. *Psychopharmacology (Berl)* 136(2): 148-152.

Schellenberg R, Sauer S, and Dimpfel W (1998). Pharmacodynamic effects of two different hypericum extracts in healthy volunteers measured by quantitative EEG. *Pharmacopsychiatry* 31(Suppl. 1): 44-53.

Sharpley AL, McGavin CL, Whale R, and Cowen PJ (1998). Antidepressant-like effect of *Hypericum perforatum* (St John's wort) on the sleep polysomnogram. *Psychopharmacology (Berl)* 39(3): 286-287.

Spinella M and Eaton LA (2002). Hypomania induced by herbal and pharmaceutical psychotropic medicines following mild traumatic brain injury. *Brain Inj* 16(4): 359-367.

Srivastava KC (1986). Isolation and effects of some ginger components of platelet aggregation and eicosanoid biosynthesis. *Prostaglandins Leukot Med* 25(2-3): 187-198.

Staffeldt B, Kerb R, Brockmoller J, Ploch M, and Roots I (1994). Pharmacokinetics of hypericin and pseudohypericin after oral intake of the *Hypericum perforatum* extract LI 160 in healthy volunteers. *J Geriatr Psychiatry Neurol* 7(Suppl. 1): S47-S53.

Thiede HM and Walper A (1994). Inhibition of MAO and COMT by hypericum extracts and hypericin. *J Geriatr Psychiatry Neurol* 7(Suppl. 1): S54-S56.

Tyler VE (1994). *Herbs of choice*. Binghamton, NY: The Haworth Press.

Vorbach EU, Hubner WD, and Arnoldt KH (1994). Effectiveness and tolerance of the hypericum extract LI 160 in comparison with imipramine: Randomized double-blind study with 135 outpatients. *J Geriatr Psychiatry Neurol* 7(Suppl. 1): S19-S23.

Wagner H and Bladt S (1994). Pharmaceutical quality of hypericum extracts. *J Geriatr Psychiatry Neurol* 7(Suppl. 1): S65-S68.

Walker DJ and Zacny JP (1998). Subjective, psychomotor, and analgesic effects of oral codeine and morphine in healthy volunteers. *Psychopharmacology (Berl)* 140(2): 191-201.

White HL, Scates PW, and Cooper BR (1996). Extracts of *Ginkgo biloba* leaves inhibit monoamine oxidase. *Life Sci* 58(16): 1315-1321.

Yue QY, Bergquist C, and Gerden B (2000). Safety of St John's wort. *Lancet* 355(9203): 576-577.

Chapter 8

Adams IB, Compton DR, and Martin BR (1998). Assessment of anandamide interaction with the cannabinoid brain receptor: SR 141716A antagonism studies in mice and autoradiographic analysis of receptor binding in rat brain. *J Pharmacol Exp Ther* 284(3): 1209-1217.

Ahlijanian MK and Takemori AE (1986). The effect of chronic administration of caffeine on morphine-induced analgesia, tolerance and dependence in mice. *Eur J Pharmacol* 120(1): 25-32.

Barsby RW, Salan U, Knight DW, and Hoult JR (1993). Feverfew and vascular smooth muscle: Extracts from fresh and dried plants show opposing pharmacological profiles, dependent upon sesquiterpene lactone content. *Planta Med* 59(1): 20-25.

Basbaum AI and Fields HL (1984). Endogenous pain control systems: Brainstem spinal pathways and endorphin circuitry. *Ann Rev Neurosci* 7: 309-338.

Belay ED, Bresee JS, Holman RC, Khan AS, Shahriari A, and Schonberger LB (1999). Reye's syndrome in the United States from 1981 through 1997. *N Engl J Med* 340(18): 1377-1382.

Bloom AS and Dewey WL (1978). A comparison of some pharmacological actions of morphine and delta9-tetrahydrocannabinol in the mouse. *Psychopharmacology (Berl)* 57(3): 243-248.

Capasso F (1986). The effect of an aqueous extract of *Tanacetum parthenium* L. on arachidonic acid metabolism by rat peritoneal leucocytes. *J Pharm Pharmacol* 38(1): 71-72.

Cassella G, Wu AH, Shaw BR, and Hill DW (1997). The analysis of thebaine in urine for the detection of poppy seed consumption. *J Anal Toxicol* 21(5): 376-383.

Catterall WA and Mackie K (1996). Local anesthetics. In Hardman JG, Limbird LE, Molinoff PB, Ruddon RW, and Goodman Gilman A (Eds.), *Goodman and Gilman's the pharmacological basis of therapeutics* (pp. 331-348). New York: McGraw-Hill.

Clark GC (1989). Comparison of the inhalation toxicity of kretek (clove cigarette) smoke with that of American cigarette smoke. I. One day exposure. *Arch Toxicol* 63(1): 1-6.

Council on Scientific Affairs (1988). Evaluation of the health hazard of clove cigarettes. *JAMA* 260(24): 3641-3644.

Damaj MI, Patrick GS, Creasy KR, and Martin BR (1997). Pharmacology of lobeline, a nicotinic receptor ligand. *J Pharmacol Exp Ther* 282(1): 410-419.

Della Bella D, Carenzi A, Frigeni V, Reggiani A, and Zambon A (1985). Involvement of monoaminergic and peptidergic components in cathinone-induced analgesia. *Eur J Pharmacol* 114(2): 231-234.

Di Chiara G and North RA (1992). Neurobiology of opiate abuse. *Trends Pharmacol Sci* 13(5): 185-193.

Dolara P, Luceri C, Ghelardini C, Monserrat C, Aiolli S, Luceri F, Lodovici M, Menichetti S, and Romanelli MN (1996). Analgesic effects of myrrh. *Nature* 379(6560): 29.

Farrell M (1994). Opiate withdrawal. *Addiction* 89(11): 1471-1475.

Ferreira SH, Moncada S, and Vane JR ([1973] 1997). Prostaglandins and the mechanism of analgesia produced by aspirin-like drugs. *Br J Pharmacol* 120(4 Suppl.): 401-412.

Fields HL, Heinricher MM, and Mason P (1991). Neurotransmitters in nociceptive modulatory circuits. *Ann Rev Neurosci* 14: 219-245.

Gilbert PE (1981). A comparison of THC, nantradol, nabilone, and morphine in the chronic spinal dog. *J Clin Pharmacol* 21(8-9 Suppl): 311S-319S.

Groenewegen WA and Heptinstall S (1990). A comparison of the effects of an extract of feverfew and parthenolide, a component of feverfew, on human platelet activity in vitro. *J Pharm Pharmacol* 42(8): 553-557.

Grossman A (1983). Brain opiates and neuroendocrine function. *Clin Endocrinol Metab* 12(3): 725-746.

Guidotti TL, Laing L, and Prakash UB (1989). Clove cigarettes: The basis for concern regarding health effects. *West J Med* 151(2): 220-228.

Hanks GW, O'Neill WM, Simpson P, and Wesnes K (1995). The cognitive and psychomotor effects of opioid analgesics. II. A randomized controlled trial of single doses of morphine, lorazepam and placebo in healthy subjects. *Eur J Clin Pharmacol* 48(6): 455-460.

Houtsmuller EJ, Walsh SL, Schuh KJ, Johnson RE, Stitzer ML, and Bigelow GE (1998). Dose-response analysis of opioid cross-tolerance and withdrawal suppression during LAAM maintenance. *J Pharmacol Exp Ther* 285(2): 387-396.

Hulse GK, Milne E, English DR, and Holman CD (1997). The relationship between maternal use of heroin and methadone and infant birth weight. *Addiction* 92(11): 1571-1579.

Insel PA (1996). Analgesic-antipyrectic and anti-inflammatory agents and drugs employed in the treatment of gout. In Hardman JG, Limbird LE, Molinoff PB, Ruddon RW, and Goodman Gilman A (Eds.), *Goodman and Gilman's the pharmacological basis of therapeutics* (pp. 617-658). New York: McGraw-Hill.

Jain NK and Kulkarni SK (1999). Antinociceptive and anti-inflammatory effects of *Tanacetum parthenium* L. extract in mice and rats. *J Ethnopharmacol* 68(1-3): 251-259.

Lane BW, Ellenhorn MJ, Hulbert TV, and McCarron M (1991). Clove oil ingestion in an infant. *Hum Exp Toxicol* 10(4): 291-294.

Laska EM, Sunshine A, Zighelboim I, Roure C, Marrero I, Wanderling J, and Olson N (1983). Effect of caffeine on acetaminophen analgesia. *Clin Pharmacol Ther* 33(4): 498-509.

Lindvall O and Björklund A (1974). The organization of the ascending catecholamine neuron systems in the rat brain as revealed by the glyoxylic acid fluorescence method. *Acta Phys Scand Suppl* 412: 1-48.

Malec D and Michalska E (1988). The effect of methylxanthines on morphine analgesia in mice and rats. *Pol J Pharmacol Pharm* 40(3): 223-232.

Markowitz K, Moynihan M, Liu M, and Kim S (1992). Biologic properties of eugenol and zinc oxide-eugenol: A clinically oriented review. *Oral Surg Oral Med Oral Pathol* 73(6): 729-737.

Mascolo N, Jain R, Jain SC, and Capasso F (1989). Ethnopharmacologic investigation of ginger *(Zingiber officinale). J Ethnopharmacol* (1-2): 129-140.

Meadway C, George S, and Braithwaite R (1998). Opiate concentrations following the ingestion of poppy seed products—Evidence for "the poppy seed defence." *Forensic Sci Int* 96(1): 29-38.

Milhorn HT Jr. (1992). Pharmacologic management of acute abstinence syndromes. *Am Fam Physician* 45(1): 231-239.

Mulé SJ and Casella GA (1988). Rendering the "poppy-seed defense" defenseless: Identification of 6-monoacetylmorphine in urine by gas chromatography/mass spectroscopy. *Clin Chem* 34(7): 1427-1430.

Murphy JJ, Heptinstall S, and Mitchell JR (1998). Randomised double-blind placebo-controlled trial of feverfew in migraine prevention. *Lancet* 2(8604): 189-192.

Musto D (1991). Opium, cocaine, and marijuana in American history. *Sci Am* 7: 40-47.

Noyes R Jr., Brunk SF, Avery DAH, and Canter AC (1975). The analgesic properties of delta-9-tetrahydrocannabinol and codeine. *Clinical Pharmacol Ther* 18(1): 84-89.

Onogi T, Minami M, Kuraishi Y, and Satoh M (1992). Capsaicin-like effect of (6)-shogaol on substance P-containing primary afferents of rats: A possible mechanism of its analgesic action. *Neuropharmacology* 31(11): 1165-1169.

Orlowski JP (1999). Whatever happened to Reye's syndrome? Did it ever really exist? *Crit Care Med* 27(8): 1582-1587.

Pasternak GW and Standifer KM (1995). Mapping of opioid receptors using antisense oligodeoxynucleotides: Correlating their molecular biology and pharmacology. *Trends Pharmacol Sci* 16(10): 344-350.

Pattrick M, Heptinstall S, and Doherty M (1989). Feverfew in rheumatoid arthritis: A double blind, placebo controlled study. *Ann Rheum Dis* 48(7): 547-549.

Pertovaara A and Tukeva T (1990). Cocaine: Effect on spinal projection neurons in the rat. *Brain Res Bull* 25(1): 1-6.

Phillips S, Ruggier R, and Hutchinson SE (1993). *Zingiber officinale* (ginger)—An antiemetic for day case surgery. *Anaesthesia* 48(8): 715-717.

Price DD (1988). *Psychological and neural mechanisms of pain.* New York: Raven Press.

Pugh WJ and Sambo K (1988). Prostaglandin synthetase inhibitors in feverfew. *J Pharm Pharmacol* 40(10): 743-745.

Reisine T and Bell GI (1993). Molecular biology of opioid receptors. *Trends Neurosci* 16(12): 506-510.

Robbers JE, Speedie MK, and Tyler VE (1996). *Pharmacognosy and pharmacobiotechnology*. Baltimore: Williams and Wilkins.

Rossi GC, Pasternak GW, and Bodnar RJ (1993). Synergistic brainstem interactions for morphine analgesia. *Brain Res* 624: 171-180.

Rudgley R (1993). *Essential substances*. New York: Kodansha International.

Saper CB and Breder CD (1994). The neurologic basis of fever. *N Engl J Med* 330(26): 1880-1886.

Sawynok J (1995). Pharmacological rationale for the clinical use of caffeine. *Drugs* 49(1): 37-50.

Scoville WL (1912). Note on capsicum. *J Am Pharm Assoc* 1: 453.

Sell AB and Carlini EA (1976). Anesthetic action of methyleugenol and other eugenol derivatives. *Pharmacology* 14(4): 367-377.

Sharma JN, Srivastava KC, and Gan EK (1994). Suppressive effects of eugenol and ginger oil on arthritic rats. *Pharmacology* 49(5): 314-318.

Smith PB and Martin BR (1992). Spinal mechanisms of delta 9-tetrahydrocannabinol-induced analgesia. *Brain Res* 578(1-2): 8-12.

Spinella M and Bodnar RJ (1996). Excitatory amino acid antagonists in the rostroventral medulla inhibit mesencephalic morphine analgesia in rats. *Pain* 64: 545-552.

Spinella M, Schaeffer LA, and Bodnar RJ (1997). Ventral medullary mediation of mesencephalic morphine analgesia by muscarinic and nicotinic cholinergic receptor antagonists in rats. *Analgesia* 3: 119-130.

Sumner H, Salan U, Knight DW, and Hoult JR (1992). Inhibition of 5-lipoxygenase and cyclo-oxygenase in leukocytes by feverfew: Involvement of sesquiterpene lactones and other components. *Biochem Pharmacol* 43(11): 2313-2320.

Szallasi A and Blumberg PM (1999). Vanilloid (capsaicin) receptors and mechanisms. *Pharmacol Rev* 51(2): 159-212.

Tyler VE (1994). *Herbs of choice: The therapeutic use of phytomedicinals*. Binghamton, NY: The Haworth Press.

Vogler BK, Pittler MH, and Ernst E (1998). Feverfew as a preventive treatment for migraine: A systematic review. *Cephalalgia* 18(10): 704-708.

Weber JT, O'Connor MF, Hayataka K, Colson N, Medora R, Russo EB, and Parker KK (1997). Activity of parthenolide at 5HT2A receptors. *J Nat Prod* 60(6): 651-653.

Young ER and MacKenzie TA (1992). The pharmacology of local anesthetics—A review of the literature. *J Can Dent Assoc* 58(1): 34-42.

Chapter 9

Abraham HD (1983). Visual phenomenology of the LSD flashback. *Arch Gen Psychiatry* 40(8): 884-889.

Abraham HD, Aldridge AM, and Gogia P (1996). The psychopharmacology of hallucinogens. *Neuropsychopharmacology* 14(4): 285-298.

Culver CM and King FW (1974). Neuropsychological assessment of undergraduate marijuana and LSD users. *Arch Gen Psychiatry* 31: 707-711.

Davis M (1987). Mescaline: Excitatory effects on acoustic startle are blocked by serotonin$_2$ antagonists. *Psychopharmacology (Berl)* 93(3): 286-291.

Flicker C, Serby M, and Ferris SH (1990). Scopolamine effects on memory, language, visuospatial praxis and psychomotor speed. *Psychopharmacology (Berl)* 100(2): 243-250.

Glennon RA (1990). Do classical hallucinogens act as 5-HT$_2$ agonists or antagonists? *Neuropsychopharmacology* 3(5-6): 509-517.

Grinspoon L and Bakalar JB (1986). Can drugs be used to enhance the psychotherapeutic process? *Am J Psychotherapy* 40(3): 393-404.

Halpern JH and Pope HG Jr. (1999). Do hallucinogens cause residual neuropsychological toxicity? *Drug Alcohol Depend* 53(3): 247-256.

James W (1902). *The varieties of religious experience.* New York: Longmans, Green, and Co.

Lotsof HS (1995). Ibogaine in the treatment of chemical dependence disorders: Clinical perspectives. *Bulletin of the Multidisciplinary Association of Psychedelic Studies* 5: 16-27.

Matossian MK (1982). Ergot and the Salem witchcraft affair. *American Scientist* 70(4): 355-357.

Popik P and Glick SD (1996). Ibogaine, a putatively anti-addictive alkaloid. *Drugs of the Future* 21(11): 1109-1115.

Schwartz RH (1988). Mescaline: A survey. *Am Fam Physician* 37(4): 122-124.

Strassman RJ (1984). Adverse reactions to psychedelic drugs: A review of the literature. *J Nerv Ment Dis* 172: 577-595.

Strassman RJ (1995). Hallucinogenic drugs in psychiatric research and treatment: Perspectives and prospects. *J Nerv Ment Dis* 183(3): 127-138.

Chapter 10

Abel EL (1970). Marijuana and memory. *Nature* 227(263): 1151-1152.

Abel EL (1971). Marihuana and memory: Acquisition or retrieval? *Science* 173(4001): 1038-1040.

Aceto MD, Scates SM, Lowe JA, and Martin BR (1995). Cannabinoid precipitated withdrawal by the selective cannabinoid receptor antagonist, SR 141716A. *Eur J Pharmacol* 282(1-3): R1-R2.

Adams PM and Barratt ES (1975). Effect of chronic marijuana administration of stages of primate sleep-wakefulness. *Biol Psychiatry* 10(3): 315-322.

Bachman JA, Benowitz NL, Herning RI, and Jones RT (1979). Dissociation of autonomic and cognitive effects of THC in man. *Psychopharmacology (Berl)* 61(2): 171-175.

Baker D, Pryce G, Croxford JL, Brown P, Pertwee RG, Huffman JW, and Layward L (2000). Cannabinoids control spasticity and tremor in a multiple sclerosis model. *Nature* 404(6773): 84-87.

Barnett G, Licko V, and Thompson T (1985). Behavioral pharmacokinetics of marijuana. *Psychopharmacology (Berl)* 85(1): 51-56.

Beal JE, Olson R, Laubenstein L, Morales JO, Bellman P, Yangco B, Lefkowitz L, Plasse TF, and Shepard KV (1995). Dronabinol as a treatment for anorexia associated with weight loss in patients with AIDS. *J Pain Symp Manage* 10(2): 89-97.

Beal JE, Olson R, Lefkowitz L, Laubenstein L, Bellman P, Yangco B, Morales JO, Murphy R, Powderly W, Plasse TF, et al. (1997). Long-term efficacy and safety of dronabinol for acquired immunodeficiency syndrome-associated anorexia. *J Pain Symp Manage* 4(1): 7-14.

Bibra E (1995). *Plant intoxicants.* Rochester, VT: Healing Arts Press.

Boutros NN and Bowers MB (1996). Chronic substance-induced psychotic disorders: State of the literature. *J Neuropsychiatry Clin Neurosci* 8(3): 262-269.

Buonamici M, Young GA, and Khazan N (1982). Effects of acute delta 9-THC administration on EEG and EEG power spectra in the rat. *Neuropharmacology* 21(8): 825-829.

Calhoun SR, Galloway GP, and Smith DE (1998). Abuse potential of dronabinol (Marinol). *J Psychoactive Drugs* 30(2): 187-196.

Chait LD (1990). Subjective and behavioral effects of marijuana the morning after smoking. *Psychopharmacology (Berl)* 100(3): 328-333.

Chait LD and Perry JL (1994). Effects of alcohol pretreatment on human marijuana self-administration. *Psychopharmacology (Berl)* 113(3-4): 346-350.

Chan GC, Hinds TR, Impey S, and Storm DR (1998). Hippocampal neurotoxicity of delta9-tetrahydrocannabinol. *J Neurosci* 18(14): 5322-5332.

Childers SR and Breivogel CS (1998). Cannabis and endogenous cannabinoid systems. *Drug Alcohol Depend* 51(1-2): 173-187.

Chiu P, Olsen DM, Borys HK, Karler R, and Turkanis SA (1979). The influence of cannabidiol and delta 9-tetrahydrocannabinol on cobalt epilepsy in rats. *Epilepsia* 20(4): 365-375.

Clark LD, Hughes R, and Nakashima EN (1970). Behavioral effects of marihuana: Experimental studies. *Arch Gen Psychiatry* 23(3): 193-198.

Cone EJ, Johnson RE, Moore JD, and Roache JD (1986). Acute effects of smoking marijuana on hormones, subjective effects and performance in male human subjects. *Pharmacol Biochem Behav* 24(6): 1749-1754.

Consroe P, Musty R, Rein J, Tillery W, and Pertwee R (1997). The perceived effects of smoked cannabis on patients with multiple sclerosis. *Eur Neurol* 38(1): 44-48.

Dalterio S, Steger R, Peluso J, and de Paolo L (1987). Acute delta 9-tetrahydrocannabinol exposure: Effects on hypothalamic-pituitary-testicular activity in mice. *Pharmacol Biochem Behav* 26(3): 533-537.

Darley CF, Tinklenberg JR, Roth WT, Vernon S, and Kopell BS (1977). Marijuana effects on long-term memory assessment and retrieval. *Psychopharmacology (Berl)* 52(3): 239-241.

Das SK, Paria BC, Chakraborty I, and Dey SK (1995). Cannabinoid ligand-receptor signaling in the mouse uterus. *Proc Natl Acad Sci USA* 92(10): 4332-4336.

Devane WA and Axelrod J (1994). Enzymatic synthesis of anandamide, an endogenous ligand for the cannabinoid receptor, by brain membranes. *Proc Natl Acad Sci USA* 91(14): 6698-6701.

Devane WA, Hanus L, Breuer A, Pertwee RG, Stevenson LA, Griffin G, Gibson D, Mandelbaum A, Etinger A, and Mechoulam R (1992). Isolation and structure of a brain constituent that binds to the cannabinoid receptor. *Science* 258(5090): 1946-1949.

Di Marzo V (1998). 2-Arachidonoyl-glycerol as an "endocannabinoid": Limelight for a formerly neglected metabolite. *Biochem (Moscow)* 63(1): 13-21.

Diana M, Melis M, Muntoni AL, and Gessa GL (1998). Mesolimbic dopaminergic decline after cannabinoid withdrawal. *Proc Natl Acad Sci USA* 95(17): 10269-10273.

Dittrich A, Battig K, and von Zeppelin I (1973). Effects of (-) delta-9-tetrahydrocannabinol (delta-9-THC) on memory, attention, and subjective state. *Psychopharmacologia* 33: 369-366.

Domino EF (1981). Cannabinoids and the cholinergic system. *J Clin Pharmacol* 21(8-9 Suppl): 249S-255S.

Dornbush RL, Fink M, and Freedman AM (1971). Marijuana, memory, and perception. *Am J Psychiatry* 128(2): 194-197.

Drewnowski A and Grinker JA (1978). Food and water intake, meal patterns and activity of obese and lean Zucker rats following chronic and acute treatment with delta9-tetrahydrocannabinol. *Pharmacol Biochem Behav* 9(5): 619-630.

Engelsman EL (1989). Dutch policy on the management of drug-related problems. *Br J Addict* 84(2): 211-218.

Feinberg I, Jones R, Walker J, Cavness C, and Floyd T (1976). Effects of marijuana extract and tetrahydrocannabinol on electroencephalographic sleep patterns. *Clin Pharmacol Ther* 19(6): 782-794.

Fletcher JM, Page JB, Francis DJ, Copeland K, Naus MJ, Davis CM, Morris R, Krauskopf D, and Satz P (1996). Cognitive correlates of long-term cannabis use in Costa Rican men. *Arch Gen Psychiatry* 53(11): 1051-1057.

Fride E and Mechoulam R (1993). Pharmacological activity of the cannabinoid receptor agonist, anandamide, a brain constituent. *Eur J Pharmacol* 231(2): 313-314.

Fried PA (1995). Prenatal exposure to marihuana and tobacco during infancy, early and middle childhood: Effects and an attempt at synthesis. *Arch Toxicol* 17: 233-260.

Grinspoon L (1999). *Marihuana reconsidered,* Second edition. Oakland, CA: Quick American Archives.

Harclerode J (1984). Endocrine effects of marijuana in the male: Preclinical studies. *NIDA Res Monogr* 44: 46-64.

Heishman SJ, Stitzer ML, and Yingling JE (1989). Effects of tetrahydrocannabinol content on marijuana smoking behavior, subjective reports, and performance. *Pharmacol Biochem Behav* 34(1): 173-179.

Herkenham M, Lynn AB, Johnson, MR, Melvin LS, de Costa BR, and Rice KC (1991). Characterization and localization of cannabinoid receptors in the rat brain: A quantitative in vitro autoradiographic study. *J Neurosci* 11: 563-583.

Joranson DE, Ryan KM, Gilson AM, and Dahl JL (2000). Trends in medical use and abuse of opioid analgesics. *JAMA* 283(13):1710-1714.

Joy JE, Watson SJ, Benson JA (Eds.) (1999). *Marijuana and medicine: Assessing the science base.* Washington, DC: National Academy Press, Institute of Medicine.

Karler R and Turkanis SA (1981). The cannabinoids as potential antiepileptics. *J Clin Pharmacol* 21(8-9 Suppl): 437S-448S.

Kouri E, Pope HG Jr., Yurgelun-Todd D, and Gruber S (1995). Attributes of heavy vs. occasional marijuana smokers in a college population. *Biol Psychiatry* 38(7): 475-481.

Liu X, Matochik JA, Cadet JL, and London ED (1998). Smaller volume of prefrontal lobe in polysubstance abusers: A magnetic resonance imaging study. *Neuropsychopharmacology* 18(4): 243-252.

MacCoun R and Reuter P (1997). Interpreting Dutch cannabis policy: Reasoning by analogy in the legalization debate. *Science* 278(5335): 47-52.

Marks DF and MacAvoy MG (1989). Divided attention performance in cannabis users and non-users following alcohol and cannabis separately and in combination. *Psychopharmacology (Berl)* 99(3): 397-401.

Mathew RJ, Wilson WH, Coleman RE, Turkington TG, and DeGrado TR (1997). Marijuana intoxication and brain activation in marijuana smokers. *Life Sci* 60(23): 2075-2089.

Mathew RJ, Wilson WH, Humphreys D, Lowe JV, and Weithe KE (1993). Depersonalization after marijuana smoking. *Biol Psychiatry* 33(6): 431-441.

Mattes RD, Engelman K, Shaw LM, and Elsohly MA (1994). Cannabinoids and appetite stimulation. *Pharmacol Biochem Behav* 49(1): 187-195.

Mendelson JH, Mello NK, Ellingboe J, Skupny AS, Lex BW, and Griffin M (1986). Marihuana smoking suppresses luteinizing hormone in women. *J Pharmacol Exp Ther* 237(3): 862-866.

Murillo-Rodriguez E, Sanchez-Alavez M, Navarro L, Martinez-Gonzalez D, Drucker-Colin R, and Prospero-Garcia O (1998). Anandamide modulates sleep and memory in rats. *Brain Res* 812(1-2): 270-274.

Murray JB (1986). Marijuana's effects on human cognitive functions, psychomotor functions, and personality. *J Gen Psychol* 113(1): 23-55.

Musty RE and Kaback L (1995). Relationships between motivation and depression in chronic marijuana users. *Life Sci* 56(23-24): 2151-2158.

Nahas G and Latour C (1992). The human toxicity of marijuana. *Med J Aust* 156(7): 495-497.

Nelson K, Walsh D, Deeter P, and Sheehan F (1994). A phase II study of delta-9-tetrahydrocannabinol for appetite stimulation in cancer-associated anorexia. *J Palliative Care* 10(1): 14-18.

Noyes R Jr., Brunk SF, Avery DAH, and Canter AC (1975). The analgesic properties of delta-9-tetrahydrocannabinol and codeine. *Clin Pharmacol Ther* 18(1): 84-89.

O'Brien CP (1996). Drug addiction and abuse. In Hardman JG, Limbird LE, Molinoff PB, Ruddon RW, and Goodman Gilman A (Eds.), *Goodman and Gilman's the pharmacological basis of therapeutics* (pp. 557-578). New York: McGraw-Hill.

Petro DJ and Ellenberger C Jr. (1981). Treatment of human spasticity with delta 9-tetrahydrocannabinol. *J Clin Pharmacol* 21(8-9 Suppl): 413S-416S.

Pope HG Jr. and Yurgelun-Todd D (1996). The residual cognitive effects of heavy marijuana use in college students. *JAMA* 275(7): 521-527.

Rafaelsen L, Christrup H, Bech P, and Rafaelsen OJ (1973). Effects of cannabis and alcohol on psychological tests. *Nature* 242(5393): 117-118.

Rettori V, Wenger T, Snyder G, Dalterio S, and McCann SM (1988). Hypothalamic action of delta-9-tetrahydrocannabinol to inhibit the release of prolactin and growth hormone in the rat. *Neuroendocrinol* 47(6): 498-503.

Robbers JE, Speedie MK, and Tyler VE (1996). *Pharmacognosy and pharmacobiotechnology*. Baltimore: Williams and Wilkins.

Rudgley R (1993). *Essential substances*. New York: Kodansha International.

Ruh MF, Taylor JA, Howlett AC, and Welshons WV (1997). Failure of cannabinoid compounds to stimulate estrogen receptors. *Biochem Pharmacol* 53(1): 35-41.

Scallet AC (1991). Neurotoxicology of cannabis and THC: A review of chronic exposure studies in animals. *Pharmacol Biochem Behav* 40(3): 671-676.

Schultes RE and Hofman A (1992). *Plants of the gods*. Rochester, VT: Healing Arts Press.

Sherwood RA, Keating J, Kavvadia V, Greenough A, and Peters TJ (1999). Substance misuse in early pregnancy and relationship to fetal outcome. *Eur J Pediatr* 158(6): 488-492.

Tinklenberg JR, Melges FT, Hollister LE, and Gillespie HK (1970). Marijuana and immediate memory. *Nature* 226(251): 1171-1172.

Tunving K (1985). Psychiatric effects of cannabis use. *Acta Psychiatr Scand* 72: 209-217.

Varma VK, Malhotra AK, Dang R, Das K, and Nehra R (1988). Cannabis and cognitive functions: A prospective study. *Drug Alcohol Depend* 21(2): 147-152.

Vincent BJ, McQuiston DJ, Einhorn LH, Nagy CM, and Brames MJ (1983). Review of cannabinoids and their antiemetic effectiveness. *Drugs* 25(Suppl. 1): 52-62.

Walker A, Rosenberg M, and Balaban-Gil K (1999). Neurodevelopmental and neurobehavioral sequelae of selected substances of abuse and psychiatric medications in utero. *Child Adolesc Psychiatr Clin N Am* 8(4): 845-867.

Waskow IE, Olsson JE, Salzman C, and Katz MM (1970). Psychological effects of tetrahydrocannabinol. *Arch Gen Psychiatry* 22(2): 97-107.

Wilson WH, Ellinwood EH, Mathew RJ, and Johnson K (1994). Effects of marijuana on performance of a computerized cognitive-neuromotor test battery. *Psychiatry Res* 51(2): 115-125.

Chapter 11

American Psychiatric Association (2000). *Diagnostic and statistical manual of mental disorders,* Fourth edition, text revision. Washington, DC: American Psychiatric Association.

Blood AJ and Zatorre RJ (2001). Intensely pleasurable responses to music correlate with activity in brain regions implicated in reward and emotion. *Proc Natl Acad Sci U S A* 98(20): 11818-11823.

Blum K, Braverman ER, Holder JM, Lubar JF, Monastra VJ, Miller D, Lubar JO, Chen TJ, and Comings DE (2000). Reward deficiency syndrome: A biogenetic model for the diagnosis and treatment of impulsive, addictive, and compulsive behaviors. *J Psychoactive Drugs* 32 (Suppl.): i-iv, 1-112.

Chambers RA and Potenza MN (2003). Neurodevelopment, impulsivity, and adolescent gambling. *J Gambl Stud* 19(1): 53-84.

Conway KP, Kane RJ, Ball SA, Poling JC, and Rounsaville BJ (2003). Personality, substance of choice, and polysubstance involvement among substance dependent patients. *Drug Alcohol Depend* 71(1): 65-75.

Eaton SB and Konner M (1985). Paleolithic nutrition: A consideration of its nature and current implications. *N Engl J Med* 312(5): 283-289.

Eaton SB, Strassman BI, Nesse RM, Neel JV, Ewald PW, Williams GC, Weder AB, Eaton SB 3rd, Lindeberg S, Konner MJ, Mysterud I, and Cordain L (2002). Evolutionary health promotion. *Prev Med* 34(2): 109-118.

Goldstein RZ and Volkow ND (2002). Drug addiction and its underlying neurobiological basis: Neuroimaging evidence for the involvement of the frontal cortex. *Am J Psychiatry* 159(10): 1642-1652.

Howard M, McMillen C, Nower L, Elze D, Edmond T, and Bricout J (2002). Denial in addiction: Toward an integrated stage and process model—Qualitative findings. *J Psychoactive Drugs* 34(4): 371-382.

Knutson B, Fong GW, Adams CM, Varner JL, and Hommer D (2001). Dissociation of reward anticipation and outcome with event-related fMRI. *Neuroreport* 12(17): 3683-3687.

Koepp MJ, Gunn RN, Lawrence AD, Cunningham VJ, Dagher A, Jones T, Brooks DJ, Bench CJ, and Grasby PM (1998). Evidence for striatal dopamine release during a video game. *Nature* 393(6682): 266-268.

Koob GF and Le Moal M (1997). Drug abuse: Hedonic homeostatic dysregulation. *Science* 278(5335): 52-58.

Levi MS and Borne RF (2002). A review of chemical agents in the pharmacotherapy of addiction. *Curr Med Chem* 9(20): 1807-1818.

Lindsey KP, Gatley SJ, and Volkow ND (2003). Neuroimaging in drug abuse. *Curr Psychiatry Rep* 5(5): 355-361.

Martin-Soelch C, Leenders KL, Chevalley AF, Missimer J, Kunig G, Magyar S, Mino A, and Schultz W (2001). Reward mechanisms in the brain and their role in dependence: Evidence from neurophysiological and neuroimaging studies. *Brain Res Brain Res Rev* 36(2-3): 139-149.

National Institute on Drug Abuse (NIDA) (n.d.). <http://www.nida.nih.gov/>.

Nestler EJ (2002). From neurobiology to treatment: Progress against addiction. *Nat Neurosci* 5 (Suppl.): 1076-1079.

Petry NM (2001). Pathological gamblers, with and without substance use disorders, discount delayed rewards at high rates. *J Abnorm Psychol* 110(3): 482-487.

Prochaska JO and DiClemente CC (1983). Stages and processes of self-change of smoking: Toward an integrative model of change. *J Consult Clin Psychol* 51(3): 390-395.

Richards MP (2002). A brief review of the archaeological evidence for Palaeolithic and Neolithic subsistence. *Eur J Clin Nutr* 56(12): 1262-1278.

Rinn W, Desai N, Rosenblatt H, and Gastfriend DR (2002). Addiction denial and cognitive dysfunction: A preliminary investigation. *J Neuropsychiatry Clin Neurosci* 14(1): 525-527.

Shalev U, Grimm JW, and Shaham Y (2002). Neurobiology of relapse to heroin and cocaine seeking: A review. *Pharmacol Rev* 54(1): 1-42.

Sinha R (2001). How does stress increase risk of drug abuse and relapse? *Psychopharmacology (Berl)* 158(4): 343-359.

Small DM, Zatorre RJ, Dagher A, Evans AC, and Jones-Gotman M (2001). Changes in brain activity related to eating chocolate: From pleasure to aversion. *Brain* 124(Pt. 9): 1720-1733.

Spinella M (2003a). Evolutionary mismatch, neural reward circuits, and pathological gambling. *Int J Neurosci* 113(4): 503-512.

Spinella M (2003b). Relationship between drug use and prefrontal-associated traits. *Addict Biol* 8(1): 67-74.

Stewart J (2003). Stress and relapse to drug seeking: Studies in laboratory animals shed light on mechanisms and sources of long-term vulnerability. *Am J Addict* 12(1): 1-17.

Sullivan RJ and Hagen EH (2002). Psychotropic substance-seeking: Evolutionary pathology or adaptation? *Addiction* 97(4): 389-400.

Torchia M (2002). Cost-effective prevention for children living with addiction. *Behav Healthc Tomorrow* 11(6): 24-30.

Zilberman ML, Tavares H, and El-Guebaly N (2003). Relationship between craving and personality in treatment-seeking women with substance-related disorders. *BMC Psychiatry* 3(1):1.

Index

Order a copy of this book with this form or online at:
http://www.haworthpress.com/store/product.asp?sku=5308

CONCISE HANDBOOK OF PSYCHOACTIVE HERBS

Medicinal Herbs for Treating Psychological and Neurological Problems

_______in hardbound at $49.95 ((ISBN: 0-7890-1857-8)
_______in softbound at $34.95 (ISBN: 0-7890-1858-6)

Or order online and use special offer code HEC25 in the shopping cart.

COST OF BOOKS_______

POSTAGE & HANDLING_______
(US: $4.00 for first book & $1.50 for each additional book)
(Outside US: $5.00 for first book & $2.00 for each additional book)

SUBTOTAL_______

IN CANADA: ADD 7% GST_______

STATE TAX_______
(NJ, NY, OH, MN, CA, IL, IN, & SD residents, add appropriate local sales tax)

FINAL TOTAL_______
(If paying in Canadian funds, convert using the current exchange rate, UNESCO coupons welcome)

☐ **BILL ME LATER:** (Bill-me option is good on US/Canada/Mexico orders only; not good to jobbers, wholesalers, or subscription agencies.)

☐ Check here if billing address is different from shipping address and attach purchase order and billing address information.

Signature_______________

☐ **PAYMENT ENCLOSED:** $_______________

☐ **PLEASE CHARGE TO MY CREDIT CARD.**

☐ Visa ☐ MasterCard ☐ AmEx ☐ Discover
☐ Diner's Club ☐ Eurocard ☐ JCB

Account # _______________

Exp. Date_______________

Signature_______________

Prices in US dollars and subject to change without notice.

NAME_______________
INSTITUTION_______________
ADDRESS_______________
CITY_______________
STATE/ZIP_______________
COUNTRY_______________ COUNTY (NY residents only)_______________
TEL_______________ FAX_______________
E-MAIL_______________

May we use your e-mail address for confirmations and other types of information? ☐ Yes ☐ No
We appreciate receiving your e-mail address and fax number. Haworth would like to e-mail or fax special discount offers to you, as a preferred customer. **We will never share, rent, or exchange your e-mail address or fax number.** We regard such actions as an invasion of your privacy.

Order From Your Local Bookstore or Directly From

The Haworth Press, Inc.

10 Alice Street, Binghamton, New York 13904-1580 • USA
TELEPHONE: 1-800-HAWORTH (1-800-429-6784) / Outside US/Canada: (607) 722-5857
FAX: 1-800-895-0582 / Outside US/Canada: (607) 771-0012
E-mailto: orders@haworthpress.com

For orders outside US and Canada, you may wish to order through your local sales representative, distributor, or bookseller.
For information, see http://haworthpress.com/distributors

(Discounts are available for individual orders in US and Canada only, not booksellers/distributors.)

PLEASE PHOTOCOPY THIS FORM FOR YOUR PERSONAL USE.

http://www.HaworthPress.com BOF04